Revaluing Horticultural Skills

This book highlights the value and skill of horticultural work through stories of food cultivation. It examines the difficulties that arise from the perception that this type of activity is unskilled and the importance of acknowledging the expertise involved in growing food.

The book provides a rare focus on horticulture as a vital part of agri-food systems, offering a social science perspective on the sector's current and past characteristics. It presents new primary research into horticultural work and workers across UK food growing, using close attention to their abilities to highlight the depth of their knowledge and learning. This is set in the context of global agri-food regimes which press producers to seek ever more precarious labour, undermining food justice. By examining these in the context of internationally connected supply chains, it characterises injustices which recur globally and across food system labour. The conceptual argument starts from an ecological definition of skill as a social practice embedded within its socio-economic landscape, developing this perspective beyond its association with artisanal contexts. Together the empirical and conceptual materials highlight the fallacy of discourse which tends to individualise skill and the challenges around recruitment into food production. To counter this, the book proposes a more collective approach to fostering healthy skills ecosystems, reaching towards commoning through examples of horticultural communities seeking this in the meantime.

It will appeal to postgraduates, researchers and professionals interested in food systems, their workers and related topics of horticultural education, training and human resources, labour, migration and politics of injustice. It draws on perspectives from rural studies, human geography and sociology and connects with international debates in these fields. Food focused scholars and activists will find data and insights to support calls for better work in food systems.

Hannah Pitt is Senior Lecturer in Environmental Geography at Cardiff University's School of Geography and Planning, where she specialises in researching and teaching food system sustainability, with a focus on human-plant relations.

Critical Food Studies

Series editors:

Michael K. Goodman
University of Reading, UK

Colin Sage
Independent Scholar

The study of food has seldom been more pressing or prescient. From the intensifying globalisation of food, a worldwide food crisis and the continuing inequalities of its production and consumption to food's exploding media presence and its growing re-connections to places and people through 'alternative food movements', this series promotes critical explorations of contemporary food cultures and politics. Building on previous but disparate scholarship, its overall aims are to develop innovative and theoretical lenses and empirical material in order to contribute to – but also begin to more fully delineate – the confines and confluences of an agenda of critical food research and writing.

Of particular concern are original theoretical and empirical treatments of the materialisations of food politics, meanings and representations, the shifting political economies and ecologies of food production and consumption and the growing transgressions between alternative and corporatist food networks.

The Rhetorical Construction of Vegetarianism
Edited by Cristina Hanganu-Bresch

Artists and the Practice of Agriculture
Politics and Aesthetics of Food Sovereignty in Art Since 1960
Silvia Bottinelli

Food Marketing and Selling Healthy Lifestyles with Science
Transhistorical Perspectives
Edited by Lauren Alex O'Hagan and Göran Eriksson

Revaluing Horticultural Skills
The Knowledge and Labour of Growing Food
Hannah Pitt

For more information about this series, please visit: www.routledge.com/Critical-Food-Studies/book-series/CFS

Revaluing Horticultural Skills

The Knowledge and Labour of Growing Food

Hannah Pitt

Routledge
Taylor & Francis Group
LONDON AND NEW YORK

First published 2025
by Routledge
4 Park Square, Milton Park, Abingdon, Oxon OX14 4RN

and by Routledge
605 Third Avenue, New York, NY 10158

Routledge is an imprint of the Taylor & Francis Group, an informa business

British Library Cataloguing-in-Publication Data
A catalogue record for this book is available from the British Library

Library of Congress Cataloging-in-Publication Data
Names: Pitt, Hannah, author.
Title: Revaluing horticultural skills / Hannah Pitt
Description: First edition | New York, NY : Routledge, 2025 | Includes bibliographical references and index
Identifiers: LCCN 2024027713 (print) | LCCN 2024027714 (ebook) | ISBN 9781032251202 (hardback) | ISBN 9781032251219 (paperback) | ISBN 9781003281689 (ebook)
Subjects: LCSH: Agriculture--Labor productivity. | Horticultural service industry | Agricultural productivity--Social aspects.
Classification: LCC S564 .P575 2025 (print) | LCC S564 (ebook) | DDC 338.1--dc23/eng/20240916
LC record available at https://lccn.loc.gov/2024027713
LC ebook record available at https://lccn.loc.gov/2024027714

ISBN: 978-1-032-25120-2 (hbk)
ISBN: 978-1-032-25121-9 (pbk)
ISBN: 978-1-003-28168-9 (ebk)

DOI: 10.4324/9781003281689

Typeset in Times New Roman
by SPi Technologies India Pvt Ltd (Straive)

Contents

Acknowledgements

Barbora Adlerová and Rhiannon Craft worked as research assistants on projects that grew out of my fellowship, which contributed to the thinking presented here. Thanks to them for being far more capable and willing than I could dare to hope for, and always being a joy to work with. Jack Pickering assisted with early stages of the review featured in chapter four, doing a sterling job of sorting, sifting and digesting the material.

I could not have made it this far or continued to survive the vagaries of an academic career without care and inspiration from many awesome women: the Careoperative (including 'token men'), Kirstie, Agatha and other lovely colleagues in Cardiff, and of course LC, HB, Nat and Rhi, the original awesome women of 5B. I am particularly grateful that this community led me to other great women – Lydia Medland and Susanna Klassen – who became such wonderful collaborators.

Thanks to Gill Bristow and Terry Marsden for mentoring my fellowship project and always being so encouraging.

Special thanks to Colin Sage and Mike Goodman for their patience, clear steer and enthusiasm for this book.

And finally, thanks to Otis for taking me on many walks towards clearer thinking, and to Simon for letting me get on with it.

1 Introduction

The value of horticultural work

15 seconds. That's how long it takes. Tug, tug, pass, swipe, pass, swipe, toss, tug, tap, tie, snap, throw. Pull up a handful of spring onions, pass from one hand to the other, removing the outer skins along the way. Discard any too fat or thin, grab more to make a bunch the right weight. Align the roots, fasten them with a rubber band, snap off the tops. Toss the bunch onto the pile, straight onto the next handful. All in 15 seconds. It takes longer to write an account of the task than it took the picker to do it, and I've missed some details. It all appeared so efficient and easy, so fast. I can only capture this much because I filmed the picker, for a few bunches worth of a day harvesting crates and crates of spring onions ready to be washed and labelled, then trucked to the retailer.

That day over 100 pickers were in the field, all in high-vis vests, some with knee pads, dispersed around the rows, kneeling, or sitting on upturned crates alongside the plants, a bowl of bright blue rubber bands within reach. Piles of discarded onions were everywhere, soon squashed under foot or wheel, releasing that familiar pungent odour. Coloured flags marked the rows to work along; supervisors' glances moved between their team and handheld computer, making sure the numbers were on target. And when almost complete, a call of "last box" relayed around the field. The order was picked.

What features in that film is only part of the work ensuring good onions make it to the supermarket shelf on time. A day or so before, the farm team checked the forecast and judged there would be no rain, so directed the irrigation team to water the crop, otherwise the ground would be too hard for the pickers to easily tug up the onions. Too much water – irrigation and then rain – means the pickers compact the soil, and the crop comes up muddy. Days before that, managers checked which rows could be picked, to be sure they were not sprayed within two days of picking. That is only possible if the agronomist keeps an eye out for pests and diseases and manages to control them. Then the farm team coordinated with HR to make sure enough pickers would be available, and that the drivers would get them to the field on time. Go back further and there was the ground preparation, the seed sowing, the planning of the planting

DOI: 10.4324/9781003281689-1

schedule to align with the predicted orders, the forecasts of labour need and … and … and….. and they say it's unskilled work.

Looking across what happens to produce fruits and vegetables on any scale, it is clear that huge amounts of expertise are involved, with specialist knowledge coming together to coordinate people, plants and more. A complex ecosystem of knowledge is required to grow crops, much of which is misunderstood or overlooked, meaning the skills required in horticulture are vastly undervalued. Most undervalued of all is the skill of the picker and everything he did within just 15 seconds. They say it's unskilled work.

This book is about his skill, and that across the processes involved in getting those spring onions ready, and the many other fruit and vegetable crops which require similar know-how. It will show how skilled horticultural workers are, and why it matters that they are valued as such. It will tell you stories of food growing in the UK which reach overseas and have echoes globally. It will trace why so many people came to believe this kind of work is unskilled, and the problems this creates. And it will suggest our shared responsibility to value the skills of food growing.

Why we need to revalue horticultural skills

This book argues that skilled work essential to horticulture has been denied and misunderstood with implications for people doing the work, and our ability to grow enough good food into the future. Labelling jobs as low or unskilled is demeaning to the workers, and portrays them as stupid or replaceable, undermining their social standing. This deters people from seeking opportunities within horticulture, establishing a vicious cycle in which skill development lacks investment, reducing opportunities to become a food grower. But declining infrastructure for education and training is not the main source of horticulture's skills problems – the challenge of making a good income as a food grower and food-system dynamics which prevent this are the roots of shortages of trainees and workers wanting to become horticulturalists.

My wish is that bringing closer attention to what food growers do and know leaves no doubt: there is no such thing as unskilled horticultural work. By attending to the know-how of people across food growing, approaching them with a more appropriate understanding of skill, I present their work as skilled practices entailing complex knowledge development and exchange. The conceptual dimension of this task is to develop an ecological understanding of skill as situated practice, taking this beyond artisanal, craftwork contexts, where it has been most applied, into industrialised production systems, such as fields of spring onion pickers. This definition of skill counters versions which focus on an individual and the knowledge they hold in mind, requiring a socio-ecological perspective locating each worker within relations and flows extending far beyond them. I argue that this less individualised consideration of skill, locating it within a knowledge ecosystem, is essential to address the problems which limit horticulture's access to knowledge and skilled workers. Correcting limits acting on this ecosystem is a common responsibility shared by all who depend on the food horticulture produces.

There are other routes into thinking about horticultural work, and seeking to correct injustices around it, so why focus on skill? First is the need to address the problematic portrayal of horticultural work as unskilled, and how this is used to justify the treatment of workers. Second is to understand the nature of the sector's reputed skills shortages (Pitt 2021b), and how they affect food production. Many argue that future global diets should be more reliant on plant foods, making horticultural production more significant; staffing this growth and access to essential know-how are crucial. Knowledge is a vital element of resilient food systems (Anderson 2015), making access to food production know-how a priority for transformation of agri-food systems. Too often the detail of what this entails is overlooked or poorly formulated (Pitt 2021b). Attention to skill directs us towards the knowledge necessary for resilient food production. Skill also connects analysis across scales from very close practices of people handling plants out to the levels of action and influence stretching far beyond them. Skill sits at the tensions between micro and macro being both non-representational practice and subject to representation through qualifications and training standards (Patchett and Mann 2018). It is enrolled in politics around work, immigration and education in ways which do not always reflect the practices it is said to describe, as characterisations are distorted by political and social constructions. But skill is not wholly a social construction: workers need capacities to do good work (Osterman et al. 2022), so it entails observable, describable practices. In the case of food work, these practices are often hidden from the consumer's view, so a focus on knowledge helps spotlight the workers who are so important yet too often neglected or treated unjustly. Finally, skill is part of notions of good work and good lives, hence concern with deskilled occupations and alienation from skilled work as features of "bad work". Understanding the role of skill in contemporary work points to how workers' lives can become better. I examine whether "reskilling" jobs can help redress their devaluing in hierarchies which determine remuneration, conditions and social perceptions.

A note on terminology. Horticulture is one form of agriculture, often subsumed within the broader term, but sometimes considered separately. There is relatively little literature focused on UK horticulture, so I draw on material considering agriculture more generally. Where sources or data distinguish horticulture, this is the term used; otherwise, I refer to agriculture or agri-food. Not all horticulture grows food, with considerable industries and social activity focused on ornamental plants in gardens, parks and other spaces. These are not my focus, although some of the infrastructure around training and education serves this and "edible" horticulture. Nomenclature around who does horticultural work is similarly fluid, with people who work to grow fruits and vegetables variously calling themselves farmers, growers and horticulturalists. My terminology reflects how people describe themselves, otherwise using the term "grower" as an inclusive descriptor.

Skill is a contested term, as academic disciplines take contrasting views on what it is, where it is located and how it is acquired (Bryson 2017). It is used to label types of work or job, to describe abilities and learning, and used politically to categorise people. What we understand as skill, and perceptions of work as skilled or not, are influenced by racialised and gendered constructions of work and learning (Osterman et al. 2022; Rose 2005). Skill should therefore be approached from a perspective which recognises its micro and macro dimensions, attending to bodily practice whilst situating these within workplaces, recognising the politics which organise and interpret them. My definition of skill is outlined below. For now, hold in mind that I take an ecological approach, meaning a skilled person is situated within their work's social and material context, as a whole thinking-acting person. A lack of hierarchy between mind and body means distinctions between intellectual and manual work do not hold; there is no such thing as unskilled work. By horticultural skills I mean the abilities required to grow food, including hands-on plant work and other essential operations such as running a business. These skills involve things, plants and other nonhumans and are exchanged through flows impeded or aided by institutions and political systems. But this is not how everyone thinks about horticultural skills; alternative definitions of skill, or failure to carefully articulate it, create all kinds of problems.

Horticultural skills devalued

Knowing how to grow food is fundamental to human existence. So when those working in or familiar with horticulture report a skills crisis, warning of shortages of people wanting to become experts in food growing, it should be cause for concern. As detailed in Chapter 3, problems around recruitment and training in horticulture are long-standing. Reasons for this are unclear due to lack of research, and assumptions such as the widespread sentiment that horticulture suffers an image unappealing to younger generations (Pitt 2019). The roots of the problem run long and deep, fed by profound social change such as urbanisation and industrialisation, entangling with sociocultural ideas around work, learning and social status. These tend to construe work like that central to horticulture in ways which devalue it, both materially through remuneration and in social esteem. The devaluing of horticultural work is fuelled by two interwoven tendencies: agricultural exceptionalism treats work in food production as different, exempting it from social controls which otherwise protect waged workers. Second, hierarchical categories applied to work and jobs determine certain jobs to be unskilled. As detailed in the next chapter, horticultural work has suffered the worst effects of both, so workers in food growing are amongst the world's most exploited and worst rewarded (Klassen et al. 2023).

Next, I outline the basis of agricultural exceptionalism and social construction of skill, and how they combine to devalue horticultural work.

Agricultural exceptionalism and exceptionally bad jobs

There are many ways that agricultural work is unique, not least because being able to produce food underpins human life. Those working in food production epitomise "essential workers" so prominent during the Covid-19 pandemic; travelling to jobs in food growing was exempt from restrictions because the work was vital. States want to ensure food security and so intervene in agricultural markets in ways inconceivable for other industries, for example through farmer subsidy justified on grounds of public good (Lee 2014; Skogstad 1998). Agricultural exceptionalism is accepted because farming is subject to natural processes which are unpredictable or uncontrollable, threatening farmers' income. Plants do their own thing, making crops a risky business, earning special dispensation from regulation.

This exceptionalism has not resulted in significant income support for UK horticulture which was the least subsidised sector under the EU's Common Agricultural Policy. But biophysical properties which make agriculture so risky have influenced the sector's labour regime. Fruit and veg don't stay fresh long, creating an imperative of speed across its picking and handling, work which must protect easily damaged goods. The privileging of freshness has allowed horticultural employers to argue for exceptions from labour regulation (Freidberg 2009). More recently, UK producers deployed the image of crops rotting in the field to argue the imperative of access to seasonal migrant workers (Chapter 4). The seasonal cycles of fresh produce dictate employment patterns which have secured growers special status within immigration regimes which otherwise exclude "unskilled" workers.

The naturalness of horticultural production echoes through another dimension of exceptionalism. Agriculture is associated with the rural idyll – imagery of healthy lives, lived traditionally in natural landscapes. For many consumers, farming remains associated with idyllic small farms worked by family members (Getz et al. 2008; Gray 2014). In the agrarian ideal, farm people are happy, doing pleasant, healthy work; organic farms are imagined loci of ecological sustainability and social justice where farmers live good lives (Guthman 2014). The UK's pervasive rural idyll masks the nature of horticultural work (Scott and O'Reilly 2022), which involves huge numbers of waged workers toiling in conditions often far from pleasant. Where family still do the bulk of the labour, self-exploitation is normalised and livelihoods precarious (Galt 2013; Gray 2014; Guthman 2014; Weiler 2022). Non-commodified labour has long been a part of agriculture and in the 21st century includes the contributions of volunteers and trainees working without remuneration (Ekers et al. 2016). Food's exceptional status underpins acceptance of hard work for little pay.

How natural processes fundamental to food production make a difference to the political economy is part of the agrarian question long considered by

Marxists (Ekers et al. 2016; McMichael 1997). This reveals fundamental dimensions of agriculture which affect the status of its workers and suggests it is capitalist relations of production rather than anything about farming per se which exclude farmworkers from social controls which value other workers. The rise of capitalism followed by industrialisation and mass urbanisation created markets for food commodities (Akram-Lodhi and Kay 2010). Capitalism requires cheap food to keep workers fed and wages low (Gerbeau and Avallone 2016; Strauss 2013), but cheap food depresses producers' incomes, and hence they and their workers toil for low pay. Cheap food needs cheap labour. None of this is possible without the unpaid labour of social reproduction, largely done by women, keeping the labour-force fed, and caring for the future workforce (Cox and Federici 1975). Labour can remain cheap because the costs of reproduction are externalised (Gerbeau and Avallone 2016). The cheap food–cheap labour dynamic is particularly pernicious in horticulture (Chapters 2 and 3) because of workforce scale. For the agricultural labour force, these dynamics too often mean that their rights and rewards are undervalued; workers are treated cheaply (Gerbeau and Avallone 2016). This is not through occasional malpractice, or what happens when the food system breaks under peculiar pressure: cheap labour is systemic, and the system is working precisely as intended for those who benefit (Allen 2016). It is not a broken food system – the system was always fixed.

Agricultural exceptionalism is not a natural state for work centred on natural processes; farmwork has been deliberately made an exception by those who benefit. The exceptional treatment UK agriculture has had within market-led governance was achieved through historic cooperation between farmers and the state, mediated by farm unions (Lee 2014). Farmers' influence was apparent when new labour controls treated farms differently from other industries. The situation in the UK is less extreme than in the USA where the agricultural lobby wields disproportionate political power to protect its interests above that of workers (Schell 2002), for example securing exclusions from labour laws protecting collective bargaining (Mieres and McGrath 2021). The American farm lobby argued for exemption from labour regulation by perpetuating myths of farmworkers' character, for example as "natural-born wanderers who revel in the informality of agricultural work" and who would "chafe under the sort of regimentation common in more tightly regulated settings" (Schell 2002, 143). These arguments from across the Atlantic 100 years ago have a related narrative in the UK today. Employers deploy narratives of hardworking Eastern Europeans who just want to earn, so will work long hours, happy to toil hard for the season and then return home to live well on the spending power of British pounds (Scott and Rye 2021). The domestic workforce is supposedly unwilling to do the work, so without seasonal migration, there will be no British crops on supermarket shelves (Scott and O'Reilly 2022). These tropes are underpinned by stereotypes of which ethnicities suit farm work (Freidberg 2009; Holmes 2013; Scott and Rye 2021). In the post-abolition

USA, a majority of farmworkers were people of colour, so racism surely influenced resistance to labour controls (Schell 2002). But claims that farmworkers want to work in unique ways really protect employers who want flexibility to employ as suits them (Chapters 2–4). Sometimes agricultural exceptionalism is simply racism.

Although the UK's farm lobby has had a softer influence than its American counterparts, workers' voices are lacking from political discourse here (Chapter 4). Historically low union membership across food production and the absence of political representation for waged farmworkers disadvantage them in comparison with labour in other industries. The agricultural workforce may be unsuited to collective political action, having numerous individual employers who are harder to negotiate with and a dispersed workforce (Martin 2021). Trade unions struggle with 21st century labour arrangements prevalent in European horticulture, such as subcontracting and globalisation (Caruso 2017). Seasonal and migrant employees perhaps feel less incentive to organise, or struggle to integrate into civic action (Gottlieb and Joshi 2013; Rye and Scott 2018). This leaves agricultural workers less protected than those in other industries. Across most of the 20th century, agricultural wages were regulated to standardise earnings and contracting,[1] but as recently as 2016 the National Farmers Union (NFU) argued that horticultural businesses required distinct treatment to avoid being bankrupted by minimum wage increases (NFU 2016). Whilst securing limited exemption from wage controls, UK agriculture has secured special status in immigration controls through seasonal worker schemes in place for much of the post-war period (Chapters 2 and 3). Seasonal worker programmes gained support globally through perpetuating agrarian myths, which are highly questionable when applied to contemporary horticulture dominated by large businesses integrated into retail markets (Thomas 1986).

Migrant workers in Europe may be better protected than their global counterparts (Hahamovitch 2003), but EU citizenship does not always offer protection from exploitation (Palumbo and Sciurba 2015). In the UK they remain vulnerable due to their tenuous immigration status and lack of understanding regarding rights of redress (Rogaly 2008). Unscrupulous employers or recruiters exploit this precarity, perhaps by shortening working hours, or withholding pay and passports (FLEX 2021). Horticulture's reliance on migrant workers is no coincidence but a central logic of today's global food systems (Chapter 2). This shapes a workforce which is inherently disadvantaged and struggles to mobilise to challenge this. There are exceptions, such as moves to establish a new UK land workers union (SALT n.d.), and successful mobilisation of transient horticultural workers in Florida and Spain (Caruso 2017; Gottlieb and Joshi 2013). Agricultural workers have gained influence at times of crisis – either concern for national food security or when vacancies outnumber workers (Verdon 2017). When the labour pool dwindles, such as following clampdowns on Mexican immigration to the USA, farmworkers can demand more (Guthman 2014). Recent shortages in the UK pushed seasonal workers' wages upwards (Chapter 4), but public reactions to this indicate that these are widely regarded as low pay, low status jobs.

Agricultural work is not exceptional as part of some imagined rural idyll where wholesome family farms offer healthy work, as food production operates to the same neoliberal logic as other globalised commodity industries. The mythology of its exceptionalism shaped employment regimes which create exceptionally bad jobs, whilst keeping horticultural workers hidden. This is cruelly ironic given that agriculture is unique as an industry we can't live without. Its public goods should be grounds to protect those working to produce food, securing them decent rewards, when in reality horticultural workers are amongst those least rewarded by agri-food systems. One ground for treating farm work differently has been the suggestion that it employs unskilled labour. This characterisation is, and always has been, mistaken, so how did it become so widespread?

Myths of food growing as unskilled work

Historically, horticultural workers in the UK sat towards the bottom of the hierarchy of farmworkers with their work regarded as less specialised or skilled than that involving animal husbandry or specialist equipment (Chapter 3). The myth of horticultural work as undifferentiated by skill or experience also suppressed wages during the green gold rush in Californian agriculture (Thomas 1986). The region's salad production needed workers, and to justify low wages, employers portrayed the work as unskilled "which implies replaceable and therefore cheap" (Freidberg 2009: 177). Such unskilled jobs could (even!) be done by women, children and migrants, workers not seen to deserve good pay or protection (Chapter 2). Devaluing horticultural work as unskilled is part of the broader social construction of certain jobs as un- or low-skilled which justifies paying workers less and treating them as replaceable. Roles dominated by women and people of colour have been particularly susceptible to perception as unskilled, reinforcing their exploitation. Worker movements pushed to redefine skill to correct these tendencies, particularly the comparable worth movement which challenged job valuation bias which compared jobs dominated by women unfavourably with those dominated by men, resulting in pay inequality (Steinberg 1990). This highlighted socially constructed distinctions were based on racialised and gendered biases, not work content or complexity (Green 2011). Skill is also strongly associated with control of cross-border migration, with skill levels used to determine who can immigrate, making migration laws central to skill definitions (Liu-Farrer et al. 2021; Warhurst et al. 2017). Because race and gender are no longer acceptable social markers, skill is used to define people's social position and provide a basis for exclusion (Ostermann et al. 2022; Warhurst and Gatta 2017).

Supposedly unskilled work can be done by untrained people, and hence the narrative of horticultural work as low-skilled has justified reliance on temporary, casual labour (Hernández Romero 2012; Klocker et al. 2019). In practice, the speed and quality of work by those experienced in tasks such as fruit harvesting are vastly superior (Curtain et al. 2018; Klocker et al. 2019). Those familiar with horticultural operations recognise the complexity of what appear to be routine, simple tasks – these are not unskilled roles (Aldous et al. 2014). If seasonal work required no skill, then employers would not prefer experienced recruits, or look to them for beneficial knowledge (Dunn and Klocker 2017; Klocker et al. 2019). UK growers are quite concerned about declining numbers of workers completing a season, or returning annually, because experience increases productivity (Chapter 3). Attention to what horticultural labourers do shows their work to be skilled, employing complex knowledge, developed through informal learning, refined through practice (Hernández Romero 2012). Just because the job involves repetition, it is not mindless (Patchett and Mann 2018); the worker continuously perceives specific conditions and variations in plants to select the appropriate response (Klocker et al. 2019). But it remains easy to ignore this expertise because of historic, socially embedded ideas about work.

Horticultural work has been devalued because it sits at the bottom of hierarchies influencing how jobs are structured and rewarded: dualisms of manual-intellectual work, experiential-academic learning and artisanal-manufacturing labour. These are compounded by Western attitudes towards nonhumans which relegate plants to lower orders of being and intelligence. Work with plants is at the meeting point of these powerful imaginaries which combine to relegate horticultural work. But these hierarchies rest on dualistic thinking of people as knowing minds in fleshy bodies, located on one side of a divide from nature, dualisms now accepted as problematic. To get beyond them I follow Ingold's understanding of knowledge, reinterpreting skill as a more-than-human activity (1997), an ecological perspective on work which undermines hierarchies and revalues those poorly treated by them, including plants. Ingold dismantles dualisms by starting from a view of humans immersed in the world, but dualistic thinking about knowledge and work has also been undone by starting from social construction of jobs.

How we think about work has been organised around deeply embedded dualisms: blue-collar versus white-collar jobs, intellectual versus practical knowledge, academic versus vocational education and pure versus applied knowledge (Rose 2005). But as Rose demonstrates, those who specialise in "manual" or "physical" roles do not lack intellectual ability or learning. His attention was alerted to the thought in physical work through watching his mother waitress, noting her deft abilities to balance loaded plates, remember multiple orders and manage difficult customers, simultaneously and across arduous shifts. By appreciating the techniques of workers like her, Rose shows that supposedly manual or physical jobs are far from mindless, but require bodily intelligence and knowledge acquired on the job, skills typically demeaned in Western culture. He details skills in jobs such as plumbing or hairdressing, including attending to the environment and using tools, with even relatively routine tasks requiring the worker to self-monitor and then adjust their actions. In addition to these more bodily expressions of intelligence, he notes that "manual work" involves knowledge associated with more "intellectual" jobs such as memory, literacy and problem-solving. He shows manual work to entail sophisticated mental work, highlighting how jobs on either side of the supposed division between manual and intellectual work vary by degree rather than kind in how they combine bodily and conceptual knowledge. If we consider what workers do and know, it becomes impossible to consider their work as unskilled. Without biases and binaries, the only ground for categorising jobs and knowledge is to serve particular interests, by undermining some workers' status (Iskander 2021; Rose 2005). Jobs supposedly requiring sophisticated mental abilities are better rewarded and have higher status.

Stratification of jobs and work based on assumptions about their knowledge intensity has been apparent within agriculture where esteem was conferred on roles seen to employ technical knowledge and expertise. More abstract or scientific knowledge has greater social value whilst vernacular knowledge including the application of intuition is devalued, meaning those in roles perceived to rely on more embodied expertise have lower status or are regarded as unskilled (Legun et al. 2023). So agricultural work has been organised by the false binaries and politically influenced social constructions of skill that Rose identified. These ideas of skill also influenced economic and educational policy in the UK. Post-war, skill was associated with high-level qualifications, deployed in jobs requiring analytical or technical abilities, namely "the technologist, the scientist, the technician and the craftsman" (Payne 2000, 355). But pressure to re-evaluate jobs to establish gender equity revealed how thinking about skill had been influenced by conventional conceptions of male dominated work (Steinberg 1990). The false dualism of intellectual-manual work profoundly influenced thinking across education systems in North America and much of Europe. Young people are streamed into intellectual or vocational education – often along racialised or classed lines – on the assumption that different job types require different knowledge and learning, deserving different prestige and pay (Rose 2005). If unskilled work is "mindless", those destined for those jobs do not need intellectual education. Vocational education and training have been marginalised in favour of academic education and universities (OECD 2010), particularly in the UK (Batterham and Levesley 2011; Leitch Review 2006; Payne 2003). The intense expansion of higher education since the late 1990s cemented the reputation of university degrees as a source of high earnings and status (Unwin 2010). The lower social status afforded to vocational careers discourages young people and their parents from these pathways, feeding a vicious cycle of low participation and low investment (Atkins et al. 2011; OECD 2010). Despite successive UK governments espousing the importance of vocational education, it lacks social esteem and infrastructure (Atkins et al. 2011). Lack of parity with academic education represents "a divisive, classed and gendered educational system" (Atkins et al. 2011: 19). Higher education is favoured on the assumption that it creates more productive people who earn higher incomes which grow the economy (Lauder et al. 2017). Meanwhile, experiential, employment-based learning is disregarded as a source of skilled workers (Pitt 2021b). The second-class status of vocational training is apparent in horticultural education which has seen dwindling numbers of recruits and under-investment (Aldous et al. 2014; Pitt 2019). One stakeholder told me "horticulture education is fighting for its life" as low student numbers make courses unviable and shrinking budgets undermine quality.

Horticulture and other vocational careers have a devalued status which is self-reinforcing as they attract less investment and fewer recruits; social construction of jobs as "low skilled" is strengthened by educational and economic policies favouring "high skill" occupations. As Rose and others demonstrated, considering the real nature of any profession reveals hierarchical categories of work and education as unsubstantiated, resting on misrepresentation of intelligence and biases favouring jobs dominated by White Men. Studying workers regarded as unskilled led Rose to a view of intelligence beyond mind-body dualism. His appreciation of bodily intelligence, and how multiple ways of knowing interact during work, has parallels in Ingold's view of knowing through bodily immersion in the world (1997, 2000). All knowledge is practical and worldly; there is no disembodied intelligence which prefigures action in the world and so no basis for distinguishing manual and intellectual work. Rose starts with what workers do, whilst Ingold starts from what skill is; both take us beyond categorising work as skilled or not.

To summarise: the skills of horticultural work have been neglected, underestimated and undervalued through association with manual rather than intellectual work, vocational rather than academic learning, positioning them on the "wrong" side of hierarchical dualisms regarding people as mind+body, and knowledge as wholly mental. This has been reinforced by deliberate construction of certain jobs as unskilled to justify low pay and treatment of people as labour sources rather than equal citizens. These tendencies are confounded and intensified for work within agricultural settings which has historically been regarded as exceptional, leaving it free from workplace protections: workers on idyllic family farms don't need them, whilst waged labourers can't have them because the needs of the produce take precedence. Horticultural work, alongside other "manual" roles, has been undervalued because of misleading notions of skill, and deliberate neglect of the real nature of the work. A better notion of skill is required to fully appreciate what horticultural workers do.

Redefining skills – the ability to grow food

At its simplest, skill means ability gained through practice, but ideas of skill in academic debate are fluid and contested. Views on skill tend to locate it with job, or person or within social constructions, but ideally skill is understood beyond the individual by combining perspectives from psychology, social science, learning theory and political economy (Bryson 2017; Payne 2017; Warhurst et al. 2017). I follow other human geographers to regard skill as practice situated amongst multi-scalar relations, both technical and political (Patchett and Mann 2018). Counter to more orthodox interpretations of skill which treat it as the application of knowledge, through a "knowing mind" informing a "doing body", this ecological perspective unites mind and body, situating intelligence throughout person and their interactions with the environment.

This notion of skill begins with thinking and learning not as mental tasks done in the mind prior to bodily action, but throughout bodies (Ingold 1997, 2000). In Ingold's view, skill is not manual or bodily deployment of a preconceived plan or cognitively learnt rules, but rather "thought in action". It is not merely a habitual repetition of bodily techniques as seen in applying mechanical force to an object, as it entails care, dexterity and judgement (Pye 1968), requiring directed attention. Even the apparently simple, repetitive motion of striking an object with a hammer requires minute adjustments to trajectory and force, meaning the hammerer must continually perceive the hammer and hammered, correcting their motion to achieve the task (Ingold 2000, 353). Skill entails exploration and improvisation as practitioners engage attentively with their environment, noticing changing conditions and responding. Although Pye's understanding of workmanship (sic) has key differences from Ingold's approach, both emphasise that workers must exercise care and judgement as they enact knowledge. Pye says care is essential because the quality of the result is not predetermined, with the degree of risk varying across types of production (1968). Results are indefinite because conditions and materials vary, making a skilled worker one able to maintain desired results in the face of variation. Because contexts constantly change, skill cannot be learnt by rote in the form of rules to apply consistently; the learner must become able to perceive and respond to their surroundings and the objects of their work. This cannot be learnt through sharing representations of knowledge but by working alongside and communicating with more experienced practitioners (Lave and Wenger 1991). Participating alongside others, novices observe whilst engaging, being guided to attend to the environment, repeating techniques to get the feel for them, until enfolded in their body (Ingold 1997). All knowledge is practical and worldly; there is no cognitive disembodied intelligence which prefigures action in the world. It is therefore impossible to separate manual and intellectual work or to dismiss any work as unskilled.

More traditional views of skill contrast it with information which can be articulated and so made explicit, as non-linguistic or implicit knowledge

(Sigaut 1993). Skill as tacit or sub-conscious knowledge is seen to defy codification or articulation (Gertler 2003: 78). Codified or explicit knowledge is supposedly more easily communicated as information or scientific principles (Gertler 2003). This distinction is typically spatialised: tacit knowledge acquired through experience and connected to context is of the local scale, whilst explicit knowledge is more standardised and so shareable across extensive scales (Allen 2002; Morgan and Murdoch 2000; Wilbur 2014). This is why knowledge-intensive food production such as agro-ecology is seen to rely on close personal exchanges of learning between bodies doing together (Carolan 2011; Morgan and Murdoch 2000; Sporleder and Moss 2002). Technical scientific expertise is assumed universal and acquirable distant from farms where it will be applied (Fitzgerald 2003; Legun et al. 2023). Industrialisation of agriculture since the early 20th century favoured standardisation and transferability, prioritising generalisable expertise applicable across locations, over that tied to specific farms (Fitzgerald 2003). This led some to assume farmers were becoming less expert, relying more on knowledge held within inputs such as fertilisers, deferring to companies which know how to make and sell these products (Carolan 2011; Morgan and Murdoch 2000). But social research repeatedly finds that even in industrialised settings, farming relies on embodied, intuitive, and experiential knowledge (Lowe et al. 2019; Nuthall and Old 2018). Explicit knowledge only makes sense once put into practice, so "conventional" farmers apply considerable embodied know-how (Carolan 2011, 20). Farmers decide what to do through combining formalised and tacit knowledge, drawing on more universal principles and place-based know-how (Ingram 2008; Krzywoszynska 2016; Morris 2006; Oreszczyn et al. 2010). Attention to farmers' intuitive, embodied expertise dissolves apparent divisions between "skilled work" requiring inalienable expertise and "unskilled work" reliant on instruction and repetition (Legun et al. 2023). Studies of horticultural techniques such as pruning and harvesting blur this division by showing supposedly basic manual tasks to be quite skilled, with "unskilled" seasonal workers making complex decisions (Klocker et al. 2019; Legun et al. 2023). Even quite embodied knowledge is not wholly tacit or located within the worker as it features in documents and workplace materials employed in learning (Fisher and Botticello 2018). An ecological version of skill emphasises it as collaborative and situated, meaning it reaches into the world and involves others (Patchett and Mann 2018), as practitioners learn together through doing and talking about doing (Lave and Wenger 1991). Know-how is developed not only through formal education but also through practice and peer learning. On-the-job learning is central to workforce skills development, passing tacit and explicit knowledge (Lam and Marsden 2017). It is also how horticultural workers learn tasks and correct errors (Klocker et al. 2019; Hernández Romero 2012).

Following Ingold's ecological lead, the overlap between knowledge and skill becomes so significant that distinction is unnecessary; both are know-how developed by perceiving the body engaging with the environment and expert guides (Ingold 2016). Knowing is to have the capacity to perform an activity, so is inseparable from doing (Gherardi 2006). This is knowledge as situated practice, developed through and in social interaction which cannot be wholly extracted from a situation to be codified as information (Gherardi 2006, Sørensen 2009). Information recorded as "data" is equally a product of interpersonal relationships, shaped by a situation, meaning tacit and codified knowledge are not wholly distinct. The binary also dissolves in the other direction as know-how is never wholly embodied and can be communicated in writing or speaking, then transferred remotely.

I have demonstrated elsewhere that tasks such as managing a community garden, designing the site and securing funding are skilled making (Pitt 2013). Like the gardener, the project manager attends to flows through the environment, but rather than positioning soil, seed and water to foster growth, their attention lingers on immaterial flows of money and information, which their movements shape towards the desired outcome (Pitt 2013). Working with spade and soil, or computer and spreadsheets, a skilled worker anticipates how others behave and conditions towards their desired result and then moves things into place. Skill marries attention with action, and applies to all manners of contemporary work, as office and fieldworkers seek to align forces in the world, moving materials and representations into place.

This points to a final feature of skill: the involvement of materials and nonhumans, the plants, soil, water and others active in food growing. Materials always make a difference to learning (Aberton 2012; Gieser 2014), and move between spaces, taking knowledge with them (Sørensen 2009). Objects alter knowledge practices, such as complex machines which cause operators to develop new skills of working machine (Ingold 1997). Technology and human commingle, with skill distributed across them and beyond individual workplaces (Richardson and Bissell 2019). Because processes like weather and plant growth are never wholly predictable, horticulture exceeds the contrast between risky and certain production, according to which Pye assigns degrees of craftship: the more uncertain the outcome, the more skilled the producer (1968). Horticultural production always entails riskiness because lively nonhumans defy control, so growers' ability to coordinate complex life forces should be recognised as particularly skilful.

Ingold has inspired attention to skill in multiple contexts, including highly industrialised settings unlike indigenous and artisanal cultures he considers (Carr & Gibson 2016; Fisher and Botticello 2018; Gieser 2014). However, he has been criticised for silence on macro forces, not least power (Hornborg 2018). But power *can* be folded into an ecological approach to skill, for example following Gherardi to see it as what holds relationships or patterns of materials together (2006: 49). Power shapes the context in which skilled practitioners act, for example holding an organisation as a relatively stable collection of elements with conditions such as employee relations affecting workplace skills (Gertler 2003). These effects become apparent by following knowledge practices, flows and connections, from individuals out to complex, extensive systems (Gherardi 2006). The dynamics of global agri-food systems detailed in

Chapter 2 are crucial context for horticultural skills, tightly squeezing or even cutting off flows of knowledge.

Power is always a factor in understanding skill because it is such a politically laden notion. As Iskander's focus on "unskilled" migrant workers in Qatari construction deftly shows, skill is less a matter of competence than of power (2021). Workers there were categorised as unskilled not by intelligence or task complexity, but to bound political categories of people: designation as unskilled brought subjugation and curtailed rights. Categories of skill have economic, social and political effects as "representations interact with other social categories, attaching themselves to signifiers of race, gender, and class, and amplifying the social hierarchies they produce" (Iskander 2021, 9). Iskander's analysis highlights how regarding workers as unskilled leads to their political disenfranchisement and dehumanisation as non-citizens. As a social marker skill benefits from being more palatable than race or gender, whilst serving to create similar exclusions (Warhurst et al. 2017). Iskander demonstrates how capitalism extracts skill from the people who learn and practise it, making them bodies not persons, so skill becomes another form of capital: "fungible, independent of any context or persons", conveniently forgetting that it does not exist without the people who learn and practise it (2021, 12). By giving skill existence independent of the person, it is possible for some people to lack it, denying them personhood. Skills regimes, such as occupational classification determining who can immigrate to the UK, give no social, economic or civic standing to the unskilled, making them disposable and excludable (Anderson in Osterman et al. 2022). By highlighting these politics of skill, Iskander gives further reason to apply an ecological understanding. Skill carries harmful power when regarded as a capital which can exist independent of human practice and learning, as something which bodies either have or don't. Like Ingold, Iskander suggests skill is in the practice, not a thing but a flow, not an asset out there which might attach to a person, but their embodied expression of "creativity, intelligence, and will" (2021, 16). Without those who do skilled work, there is no skill.

The tendencies Iskander identifies occur elsewhere. We will see how employers in horticulture downplay their workers' abilities, and how politicians count them as numbers of labourers rather than as lives deserving full citizenship. Skills shortages are treated as deficits of abstract expertise, rather than the result of flawed systems which impede and mis-shape knowledge flows, blocking the people who want to learn growing. Media and policy discourses analysed in Chapter 4 demonstrate that definitions of skill and distinctions between high and low skill are significant in UK immigration policy, giving "highly skilled" workers the power to move. Rather arbitrary notions of skill arbitrate who deserves to migrate and stand in for notions of race (Liu-Farrer et al. 2021). Categorising roles as low- or unskilled is a highly political move, where those with power to define salaries, employment conditions or immigration status further their interests over those of workers, particularly those of marginalised race or gender. Routes into skill acquisition are also inequitable, as disempowered groups are directed into educational pathways deemed appropriate for them (Rose 2005), and migrants from marginalised countries have limited access to roles where they might upskill (Siegmann et al. 2022). Attention to such dynamics requires a version of skill which locates micro-practices centred on human bodies, within macro and meso contexts from workplace relations of the polytunnel or field up to food systems where produce moves globally.

The resulting definition of skill is:

> knowledge as socially embedded practice, enabling achievement of a goal. Knowledge and skill combine thought and practice, thoroughly bodily abilities of 'thinking through doing'. Knowledge and skills are not personal attributes, being embedded in environments which include institutional context, materials and technologies.
>
> (Pitt 2021b, p64)

Horticultural skills are the abilities required to grow food (Figure 1.1), including the know-how to grow food plants, operate an enterprise and adapt to change. These tasks are undertaken in the field, greenhouse, office, boardroom and research lab by people working across roles categorised as manual, intellectual, specialist and unskilled, including unqualified seasonal workers (Chapter 5). Skill is interchangeable with knowledge, as know-how is

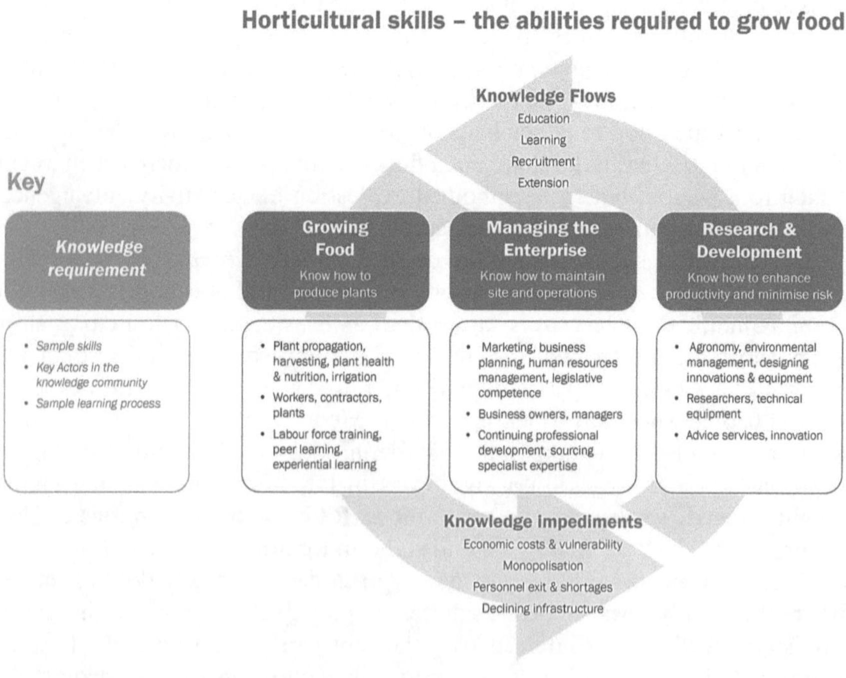

Figure 1.1 Horticultural Skills Redefined (Pitt 2021b).

distributed across and between people and others involved in food growing. Attention to how growers develop skills reveals a highly social process, enrolling materials and nonhumans, peers and experts (Chapter 6). I use the word abilities deliberately, to denote the significance of having the power to grow food. Horticultural skills can only be learnt and practised by people with the capacity and freedom to engage in food growing, which is often difficult to access and maintain in the UK (Chapter 2).

As represented in Figure 1.1, horticultural knowledge circulates through transfers involving but not reliant on codified information, plus materials and social interactions, including those with nonhumans. A considerable proportion of horticultural knowledge exchange occurs within workplaces, through learning on the job (Klocker et al. 2019; Hernández Romero 2012; Pitt 2021b). Knowledge impediments indicate power dynamics affecting skill acquisition, with the ability to learn or transfer knowledge never freely available to all. Blockages, leaks and other limits on growers' access to knowledge often arise from their limited economic capacity (Pitt 2021b). How these weaknesses affect horticultural enterprises in the UK and their knowledge ecosystems highlights skills challenges as collective problems arising from systemic pressures (Chapter 7). What is often called horticulture's skills crisis is more accurately a crisis in agri-food systems driven by cheap food economies and globalised supply chains.

The skilled spring onion picker

To reinforce this conceptual approach to skill, I want to return to the spring onion picker because their work is most unlike that traditionally regarded as skilled, and because I expect some resistance to a version of skill which sets their work alongside that of farm managers with agricultural degrees or technical expertise. Indeed, some working within horticulture deny pickers are skilled, defining skill as something other than the physical dexterity or "knack" they exhibit (Chapter 5). The picker works within an industrialised system organised to Fordist logic where labour is divided across the workforce, each worker delivering a small part of production made efficient through mechanisation and rationalisation (Fitzgerald 2003). Such production is often characterised as deskilled, contrasted with more artisanal or craft-based approaches said to allow workers greater freedom and require more expertise (Pye 1968; Sennett 2009). Deskilling theory assumes that mass production processes, automation and technology remove intelligence from the workers' domain as more intellectual tasks such as planning and conception precede processes they work on, leaving their labour to execute tasks set up by others (Vallas 1990). The assumed result is homogenised, degraded labour, far inferior to traditional craft skills (Vallas 1990). The resultant binary between craft or artisanal work and industrialised labour is another logic underpinning job hierarchies, although critical attention to recent revivals of crafting reveals a messy landscape where craft and industrial production overlap (Fox Miller

2017). In anthropology and human geography, the study of skill has favoured more artisanal settings, skewing away from automated, factory production (Carr and Gibson 2016; Fisher and Botticello 2018). But as Carr and Gibson argue, focusing on workers' practices and how all makers engage with materials reveals continuities across small-scale craft and large-scale Fordist production, as all workers share sensibilities and dispositions demonstrating careful engagement with materials (2016). Studies like theirs, and similar perspectives applied to large-scale commercial horticulture (Klocker et al. 2019), show how a notion of skill derived from non-industrialised craft settings considered by Ingold suits industrialised work, and that skill intensity does not correlate with styles or scales of production.

Three assumptions associated with deskilling have positioned tasks like spring onion picking beyond the domain of skilled work: the work is repetitive, automated, and stratified or divided. But these assumptions falter if we attend to the nature of the work from an ecological perspective. The first ground for discounting the skill of spring onion picking is the degree of repetition: the assumption that the worker merely repeats an action over and over, exhibiting considerable monotony as habitual motion is remade without applying intelligence. This is apparent in descriptions of roles in mass production horticulture as mindless, with workers engaged in tasks such as packing tomatoes "numbed by the repetitive motions" (Barndt 2008, 271). Automated, repetitive tasks are regarded as mindless because they are assumed to resemble unthinking habits – bodily rather than mental (Patchett and Mann 2018). But if intelligence is bodily, then such work is not necessarily unskilled. What appears to be repetition of the same motion over and over entails minute adjustments to how the body moves (Carr and Gibson 2016; Klocker et al. 2019). Each swing of a hammer is adjusted as the worker wielding the tool judges the position of the nail, surrounding conditions and how much force is required (Ingold 2000). Similarly, the picker attends to conditions so they can respond with the appropriate movement and move other materials into formation. The care and judgement workers apply are particularly important when working with natural materials which are highly variable and dynamic (Pye 1968). The picker feels the soil's viscosity, how deeply and firmly the onions are rooted, judging how much force will ease the onion out without snapping, noticing where there is loose skin and mud to swipe away, manipulating the bunch so roots align, judging which bulbs are on-spec, assessing the weight of the bunch and then which onion to add or remove to make target weight. Their techniques apply learnt knowledge, exercising care and judgement to ensure the desired result which is not predetermined but dependent on their dextrous action (Ingold 2000; Pye 1968). Their tools may be relatively unsophisticated but they manipulate the knife to trim roots to length, then twist the elastic band to fasten the bunch. This complex swirl of activity requires them to attend to plant, soil and equipment, then adjust their motions to achieve the desired result, making it skilled practice (Ingold 2000, 2011). The picker cannot apply rules of picking by rote because soil and onion are always different; they hold in mind the specifications of onions ordered by that day's customer and deploy bodily intelligence to form suitable bunches. The difference between a clumsy unskilled worker and one who has developed skilled dexterity is not that the latter has internalised the rules of how to complete the task, but how they attune perception and movement:

> the skilled maker who has a feel for what she is doing is one whose movement is continually and subtly responsive to the modulations of her relation with the material. Conversely, the clumsy practitioner is precisely one who implements mechanically a fixed sequence of instructions, while remaining insensitive to the evolving conditions of the task as it unfolds.
> (Ingold 2000, 357)

The novice picker does not learn by being told how to pick, but by doing it, building speed by doing it repeatedly, gaining a feel for the task. Tasks which entail repetition do not lack skill, even when more automated than spring onion picking, as detailed in a study of machinery made lace (Fisher and Botticello 2018). One reason such skills are underestimated is that proficient workers make tasks appear easy (Attewell 1990). As explored in Chapter 5, when viewing strawberry packers rapidly checking and correcting the weight of punnets on the conveyor, my guide and I marveled at their efficacy: "you just wonder how the hell they do it". Asked to explain what they are doing, they struggle (Iskander 2021, Rose 2005). Tacit knowledge disappears from view and verbalisation (Polanyi 1958), making it easier to assume workers know nothing or are not doing intelligent tasks.

A second assumption brought to notions of deskilling is that automation reduces the workers' skill as the intelligence goes into making the machine, removing it from labourers' actions. Braverman (1998) developed the concept of deskilling from Marx's ideas of industrialised labour alienating workers, as capitalism separates the value of their labour from its skill. He saw technology intensifying this by increasing the surplus value each worker can generate and denying workers knowledge of complete manufacturing processes. On Taylorist production lines, like those common across food production, workers apparently do not have to think because they specialise in part of the production, and managers or machinery designers have done the intellectual work (Busch 2019; Previtali and Fagiani 2015). Braverman saw workers as the "hands" monitored and controlled by distant "brains", leaving them without knowledge. The decline of craftwork in favour of mechanised production has been said to enable deskilling if technology intervenes in feedback between "hand and head" (Sennett 2009). Sennet did not say machinery inevitably reduces skill development, except where it prevents workers from learning by intervening in interactions between hands-on learning and mental understanding. But the mind-body dualism of this conceptualisation is problematic; it also misconstrues work in automated settings (Carr and Gibson 2016). Technology could only cleave workers from learning if knowledge travels from the hand operating machine to the mind processing information, which is not how skill works. The strawberry packers read the weight of a punnet rejected by the weighing conveyor, then select fruit to adjust the contents to the target weight, re-read the scales and pass it back to the conveyor, processing information with bodily senses and intellectual processing, all as quickly as possible and without damaging the fruit. It is a routine and repetitive task, but far from mindless. They register numeric data and judge fruit weight, at speed, and enact bodily intelligence to move fruit with care. These workers, typically classified as the least skilled in horticulture, do not enact instructions by rote, rather are engaged in constant interaction between body-brain-environment as they respond to changing variables, demonstrating skill as thinking-acting-responding (Patchett and Mann 2018). Those who work with factory machines apply somatic memory developed through practice which seems to work unconsciously, but entails dynamic work to think through what is happening, then judge what to do (Fisher and Botticello 2018).

A process powered by machinery rather than hand tools does not determine the "workmanship" (sic) required as any such distinction is impossible to sustain (Pye 1968). Automation may reduce the degree of uncertainty around the results of production, but the risk varies on a continuum rather than definitively separating machine and hand production (Pye 1968). Materials have always been part of skilled work, as tools and things made into other things (Carr and Gibson 2016). Involving more technology and more sophisticated machinery does not result in deskilling, rather the nature of the skill changes (Fisher and Botticello 2018; Rose 2005), distributing it across bodies, machine, notes and materials (Fisher and Botticello 2018; Richardson and Bissell 2019). What really matters is the context in which machine and worker interact, and who controls it (Previtali and Fagiani 2015). This brings us to the third and most significant feature of supposedly deskilled work environments: division of labour.

Strawberry packers on a partially automated production line, focused on one part of the process for a whole shift, working to the rhythm of the conveyor belt, seem to exhibit classic deskilling in Taylorist production divided into multiple discrete tasks. As Adam Smith first described, industrialised work becomes mindless (Sennett 2009), as workers repeat tasks made as efficient as possible, leaving little room for their judgement (Busch 2019). Post-war industrialised division of labour is also stratified, with roles centred on "mind" tasks seen as more valuable than those completing "body" tasks (Carr and Gibson 2016). Having dismantled the mind-body dualisms under-pinning this dimension of deskilling, something significant about the division of labour remains. The distribution of tasks and disarticulation of skill were driven by efficiency, to extract maximum production from each worker's labour, whilst increasing control over them (Busch 2019). By breaking down processes which had previously been controlled by a single artisan and dividing them between workers, power shifted to employers (Fitzgerald 2003). The Arts and Craft movement was less concerned with the introduction of machinery than with crafters' loss of control over their work through the division of labour (Pye 1968). It is the organisation of labour and how this disciplines workers rather than levels of automation which deskilling criticises: the problem with modern workplaces is not too much tech, but too little reward and protection for workers (Sennett 2009). As advanced digital technology becomes more prominent in agriculture, some regard it as a form of deskilling, but the real concern is how technology increases corporate control, reducing farmers' autonomy (Klerkx et al. 2019).

There was a feature of the strawberry packhouse which troubled me. The packhouse was a typical light-industrial unit, bland corrugated metal, blank inside and out no windows at ground level, so workers were always in artificial light. There are windows on the first-floor internal walls, opening onto the offices above, and to the small room where managers can survey the line. I visited several packhouses with these gantries; often my guide took me up to explain the operation and observe the workers. It always made me uneasy, particularly when I spotted workers discreetly raising their eyes to observe who was observing them. Being up high, in a space designed for viewing suggested a Panopticon, making me uncomfortable about being perceived to surveil the workers. Working under control and surveillance is a problematic tendency in contemporary workplaces (Previtali and Fagiani 2015). It is this which Marxist analysis of deskilling usefully highlights: alienated workers lose autonomy, the ability to control their work or to exhibit individualism (Attewell 1990). Workers in supposedly deskilled contexts are not always alienated from their work (Attewell 1990; Osterman et al. 2022; Rose 2005; Vallas 1990). It is not technology which is determinative, but employment structures and capitalist modes of production. Hence, even work in tech and creative industries – the "high skill" sectors governments favour – may not offer freedom, with many "knowledge economy" workers in bad jobs where they lack autonomy (Sonn et al. 2019). Similarly, when crafting has a profit imperative, it is less radical or pleasurable (Carr and Gibson 2016). What matters are relations between workers and employers, who benefit from workers' labour, how their work is managed and rewarded. Increased digitisation of agriculture will alter the nature of jobs, but the real concern is not the role of the new technology but how it risks intensifying existing exclusions and exploitative employment regimes. Technology is likely to increase capital accumulation and consolidation which favour those already holding power and money, and do little to benefit workers (Rotz et al. 2019).

Interpreting automated work within large-scale horticultural production as deskilled is problematic because the concept draws on false assumptions about knowledge and learning, separating mind and body, neglecting material, embodied and social dimensions of skill. Whilst there are differences between work in automated industrialised horticulture, and smaller low-tech settings, the grounds for calling the former deskilled are problematic. Deskilling idealises craftwork as the benchmark for assessing skill (Vallas 1990). It rests on a false binary between small-scale artisanal production and large-scale manufacturing, ignoring continuity in how things are made across settings (Carr and Gibson 2016). The concept has also replicated gendered thinking which over-values craftwork typically done by men (Steinberg 1990). It underestimates the complexity of tasks achieved in an automated setting, and the intelligence engaged in routinised tasks (Attewell 1990), including in contemporary mechanised production settings (Carr and Gibson 2016; Fisher and Botticello 2018; Patchett and Mann 2018). Rethinking what constitutes skill through a broader conception of intelligence and attention to work in industrialised, automated production systems brings skill back in. But deskilling is helpful for raising the question of what happens to worker autonomy within capitalist production systems and for highlighting the importance of workers' control over their labour. This is particularly significant in settings reliant on learning on the job. The spring onion picker may work for many seasons without learning much about crop production because a different team sows the seed, and others take responsibility for irrigation or pest control. The know-how required for crop production is widely distributed across an organisation, and out to specialist experts (Chapter 7), so individual workers have limited connection across the whole process. A picker can become very skilled in their task, but the division of labour curtails opportunities to broaden their knowledge; in a stratified employment structure, they lack control over which tasks they do, constraining their potential to learn-by-doing and ability to become a horticulturalist in their own right. Plus it is unlikely they could accumulate sufficient income to acquire the means to grow crops commercially themselves. It is not any deficiency of intelligence or ability that distinguishes pickers from other workers in the production system, but their freedom to explore and apply further know-how.

Other implications of industrialised production's highly specialised division of labour are that coordination across tasks and specialisms becomes vital horticultural know-how. In industrialised growing, success depends on how well plants, people, materials and markets are coordinated (Chapter 5). Any farmer is reliant on the plant's know-how to grow a crop (Ingold 2000, 2011), so even an owner-producer cannot contain all the skills horticultural production needs. The owner is never the sole knowledgeable actor as they draw on learning from a community of practice and rely on the plant's expertise to grow. Across production scales, skill is dispersed, therefore deindividualising how it is understood; growing crops is not an individual achievement, so resilience must come through cultivating diverse skilled actors and connections between them. Horticulture highlights that although treated as such, skill is not a fungible human capital as it does not exist independent of practice, or rest within an individual, rather is enacted in practice, through relations (Iskander 2021). Applying this ecological view on skill to workplaces with highly divided and stratified labour brings a political edge by revealing how power inequalities shape knowledge flows. How much freedom does each worker have to practise various parts of crop production, or to engage with more experienced growers so they can learn more? Who decides which element of production each worker is involved in, and how are assumptions about their role socially constructed to ascribe them different values? Whilst growers who own their production may not defer control to others, they still suffer limited agency due to pressures from supply chains which squeeze incomes and time, so they struggle to learn. For new entrants, the squeeze is on their power to acquire the means of production, especially land. There is greater continuity across scales and modes of horticulture than may be expected, with everyone's skill – the ability to grow food – shaped by political economic reality, and neither scale more skilled than the other.

My ecological definition of skill is deliberately broad, making all jobs skilled, so cannot differentiate jobs' skill levels or the pay they deserve. Instead of skill being a way to categorise labour, skill is the doing of labour (Richardson and Bissell 2019). This flattening of differences between kinds of work may concern some. Critics may suggest that regarding all jobs as skilled prevents attention to tackling bad jobs (Payne 2017). But a focus on skill can have exactly that effect as governments find it more palatable to invest in skills initiatives than to address inequalities harming workers such as company ownership or employee relations (Keep and Mayhew 2010). Skills development might be offered as consolation for the lack of more fundamental employment reforms or claimed by employers to improve things for workers without enhancing their autonomy or job quality (Payne 2000).

Nor will widening the category of jobs counted as skilled necessarily increase workers' pay and esteem. The comparable worth movement has achieved recognition for skills associated with women (Steinberg 1990), suggesting no harm in increasing public understanding of horticultural workers' abilities. Let's celebrate the remarkable 15-second achievement of the spring onion picker! But we must be cautious about revaluing his work based on skill level. Skill-based categorisation is risky because construction of skill levels has been so entwined with racialised and gendered hierarchies, and racial capitalism (Osterman et al. 2022; Raghuram 2021; Steinberg 1990), including migration policies which let skill proxy for race as exclusion criteria (Liu-Farrer et al. 2021). These constructions of skill seem uncontroversial and sneak hierarchies of race, gender and class in through the back door (Iskander 2021). Recategorising workers as skilled is therefore insufficient to enhance their standing and leaves in place skill hierarchies which enable political disenfranchisement (Iskander 2021, 256). Taking the bête noir of neoliberal employers, Iskander highlights that their much criticised strategies for controlling workers are only indirectly associated with the dominance of "unskilled" roles:

> the battle Amazon and others are setting up is not primarily over the (mis)appraisal of skill and the devaluing of certain social groups through biased skill assessments, but rather it is over who should have the political right to the aspects of personhood that skill and learning imply: the right to bodily autonomy and integrity, the right to voice and agency, the right to discernment and creativity.
>
> (in Osterman et al. 2022, 1362)

To ensure that workers have good, rewarding work, it is not the complexity of the task or their level of skill which matters so much as how they are regarded as people. Revaluing the skill of work like that in horticulture is not just about regrading jobs, but recognising them as vital people, deserving respect and rights.

Deskilling should draw attention less to task complexity, or how repetitive and automated work is, than to workers' loss of control and autonomy. Focus less on the packing conveyor and more on the observation window. Similarly, revaluing horticultural skills is about more than recognising the complexity of the work, and sophistication of learning. Rather it is about how workers are organised and treated, who owns their labour and whether they have control over their work. Skill must be analysed in contexts in which people work, considering how work is organised and rewarded. As argued in Chapter 8, task complexity and variety may matter less than workers having a stake in their employment. Recognising this requires thinking about skill ecologically, not as free-floating capacity, but as bodily practices of knowing, doing and learning food growing situated amongst relations which are hierarchically organised, stretching from close-up to globally distant. There is one final form of hierarchy to address to understand how horticultural work became devalued.

Mindless work with mindless plants

This chapter has highlighted how jobs regarded as "manual labour", entailing physical rather than mental work, have been demeaned and under-rewarded. Horticulture is far from the only industry bedeviled by misunderstanding of its work and workers. But there is something I believe confounds how horticultural work is conceived, heightening its devaluation. Horticulture is the art and science of growing plants, and plants have their own problems when it comes to how we value them. Skill-based categorisations of jobs common in Europe and North America are founded on false binaries between manual and intellectual work derived from mind-body dualism. As Ingold and others demonstrate, these divisions rive humans from animals, nature from culture, enabling humanistic chauvinism which assumes we are better than other kinds of beings. These logics also imposed hierarchies on life forms in which flora are lower order, less sophisticated or morally worthy beings. Ontologies which enable disregard for manual work have also led much Western thinking to overlook the complexity and value of plants, tendencies meeting in attitudes to horticultural work.

Through Western philosophy since ancient Greece, flora was regarded as a passive resource, excluded from respect paid to humans and animals for being wholly different, lacking sensitivity and intellect (Hall 2011). Across philosophies, including those influencing botanical science, plants are less significant than animals, which are in turn lesser than humans, so are systematically devalued as the lowest realm of beings (Hall 2011) Perhaps worse, people often fail to attend to plants in daily life, a phenomenon labelled plant blindness

(Wandersee and Schussler 1999). This visual-cognitive inability to notice plants combines with a lack of appreciation of their importance, resulting in a "misguided anthropocentric ranking of plants as inferior to animals and thus, as unworthy of consideration" (Wandersee and Schussler 1999, p82). Where plants are noticed, it was more often as collectives – a forested landscape, for example – than individual subjects (Head and Atchison 2009). Of course, people have always engaged with plants as significant individuals, and acted towards them with deep appreciation, with many ontologies appreciating plants as active, worthy subjects (Kimmerer 2013).

Given the reach of thinking of plants as lesser, or tendencies to not consider them very much, it is easy to see how plant work was underestimated and misunderstood. Western thinking has been dominated by the idea of plants as passive, or insensitive because we failed to attend to how plants are, so missed evidence of them being attentive and cooperative (Gagliano 2013). If plants are stupid, it easily follows that people who work with them need not be very smart either. Historic perspectives featured in Chapter 3 indicate that the degree of esteem awarded to farmworkers depended on perceived intelligence of the species they specialised with: horse handlers were of higher status than cattle herders who outranked shepherds. And what is dumber than sheep? Thinking of plants as unthinking is not the only reason horticultural work was regarded as unskilled; its suitability for women's and children's part-time agricultural work was also significant. But failure to recognise plants as complex actors cannot have helped.

Plant scientists now celebrate plants' abilities to respond to stimuli in sophisticated ways, to meet their needs drawing parallels with human intelligence (Mancuso et al. 2015; Trewavas 2003). Specialists describe what plants do in terms more typically assigned to animals, such as memory (Gagliano et al. 2014) and senses (Chamovitz 2013). There are philosophical moves to regard them as persons (Hall 2011), in part inspired by cultures in which they have always been such (Bawaka Country et al. 2016; Kimmerer 2013). Starting from a plant's perspective, it becomes clearer that:

> plants are up to stuff, that they have a kind of agency, a kind of intentionality, and their own way of getting interested and involved in the thicket of entanglements that constitute their multispecies ecologies.
>
> (Myers 2015, 62)

Closer attention to their being suggests that human blindness to how smart plants are arises from our failure to appreciate their difference, and the difficulty of knowing beings so distinct from us (Marder 2013). It is not that plants are stupid, just that we've been too stupid to notice their intelligence.

Recognised as intelligent beings worthy of regard, plants become active features of multispecies lives, entangled with humans. Life has always been such, but we've been quite good at thinking otherwise (Tsing 2012). A flatter ontology brings plants alongside animals, including humans, all actively engaged in processes such as food growing. Farming and gardening are not done to plants present as inert materials, as nonhumans make a difference through their labour (Palmer 2021). As actors in capitalism, plants "conduct various forms of work crucial to its maintenance and growth" (Ernwein et al. 2021, 22). The work humans do to grow abundant crops must understand plant responses to the environment and how to modify them (Dixon 2016). So growers coordinate with what plants are up to, working with their particularities (Brice 2021). Unlike those whose work engages with other humans, plant workers are always negotiating plants' inherent unknowability and unpredictability. Far from working with stupid creatures which require little intelligence to manage, plant work is highly sophisticated and takes special skills of understanding vegetal life.

Plants are knowledgeable actors contributing to crop production: nonhuman organisms perceive change and respond, shaping their motion and directing the movement of others to achieve a purpose (Ingold 2011). Plants attend to their environment and act in response, moving things towards their goals of growth and reproduction. A plant "knows" where light is and reaches towards it (Chamovitz 2013); roots grow towards water (Fogg 1963). Like the worker swinging a hammer, or picker pulling spring onions, the plant can't simply repeat the same motion to get the desired result as contexts change – light fades, water moves. Human gardeners sense and work with these movements, weaving together a "field of forces" (Ingold 2000: 342). Their skill is knowing what conditions encourage materials to combine, following seed, soil, water, attending and responding to increase the chance of plant growth. Like the skilled makers of woven textiles, growers combine perception and motion amidst a world alive with movement, engaging with material flows to achieve their goal by improvising in response to change (Ingold 2011). On this basis, Ingold establishes continuity between growing and making. The skill of making is to gather others into fruitful arrangements establishing the conditions for desired changes to occur (Ingold 2000: 86). One makes a crop by placing seed in soil, bringing water and light keeping pests away (Chapter 5). Growers shape the environment so the speed and course of further movements are altered – roots will grow through soil. With experience, they are better able to notice environmental cues and what is happening with a particular plant, then judge how to react (Ingold 2000: 415). But the most they can do as "fellow participants" with plants is establish conditions allowing growth to occur by attending to worldly motion, and directing its flows (Ingold 2000: 87 and 2011: 6).

By narrowing the gap between growing and making, Ingold increases the intelligence attributed to plants and growers, for makers of crops need to understand beings quite unlike people, with particularly opaque and unwieldy tendencies. Across my conversations with growers, they often admitted ignorance of what a crop is up to, acknowledging limits to their control. Plants' agency makes the outcome of horticulture highly uncertain, a facet which, according to categorisations of workers' abilities which valorise risk (Pye 1968), makes those who manage to achieve desirable outcomes amidst such uncertainty very able workers. Plants make horticulture a very risky business, and this makes its jobs and workers exceptionally skilled.

Researching horticultural skills

The insights into UK horticulture reported here come from the research project *Knowing to Grow: Increasing the resilience of plant-centred food production skills*, 2018–2022 supported by a Sêr Cymru II Fellowship part-funded by Cardiff University and the European Regional Development Fund through the Welsh Government. The project received ethical approval from Cardiff University School of Geography and Planning Research Ethics Committee. Methods included semi-structured interviews, participatory workshops, participant observation and go-along interviews at growing sites. Data collection was interrupted by the coronavirus (Covid-19) pandemic, which reduced on-site fieldwork, meaning greater reliance on remote methods (online interviews and workshops). I interviewed 51 individuals representing 16 growers and 17 associated organisations; a selection of these and additional recruits participated in workshops. All data was processed in NVivo to enable thematic analysis. Emerging findings were periodically explored with stakeholders, including through my participation in related networks and events.[2]

I aimed to cover a range of growers, reflecting variations in scale, supply chain and growing system, seeking participants working across growers' organisational structures, and from a range of perspectives on their growing operations including specialist suppliers and advisors. Growers were asked to recommend stakeholders, with other organisations approached based on my knowledge of the sector. All participants and their businesses or farms have been anonymised, with quotes attributed to individuals by a code in which letters indicate the participant's organisation or business type:

- S – smaller growers
- L – larger growers
- CEA – Controlled Environment Agriculture growers
- Su – suppliers of specialist inputs or services
- Sh – stakeholder organisations

Organisations in each category are distinguished by their number (L1, L2, etc.) with participants representing the same organisation allocated a second digit (S1.2, S1.3, etc.). All names in vignettes are pseudonyms.

Analysis suggested it was most insightful to distinguish between larger and smaller growers as operational scale is a key determinant of knowledge ecosystems. There is variation within these categories – not least between organic and conventional growers – but sufficiently common issues associated with scale. Smaller growers include those owned and run by one or two individuals, a family or community organisation, with fewer than ten workers, tending to sell to local supply chains and operate to agro-ecological principles. Larger growers include businesses with multiple production sites, 100+ workers, likely to supply large national retailers and operate in-house packing.

I want to acknowledge that the voices and lives of horticultural workers, particularly the seasonal migrants central to UK horticulture, are not as prominent as they should be. I am aware of the flawed irony that I write about the need to revalue their work, and highlight how often their voices are not heard, without centering their words. There are methodological and practical reasons for this, related to the research conception and constraints of doing much of it during a pandemic. I never intended to focus solely on their role or skill as I sought to reflect the breadth of knowledge flowing around food growing. But I should have done more to give them the profile and humanity they are often denied. There are academics and organisations that do so, and I urge readers to engage with their many materials which I refer to and appreciate.

Conclusion

This chapter laid the conceptual groundwork for the empirical discussions which follow, showing how horticultural work and workers became devalued through typical Western thinking about work and learning. Binaries which falsely divide manual and intellectual abilities, demeaning the former, have stratified jobs into hierarchies shaped by gendered and racially biased notions of who suits what work, and the level of reward they deserve. Revaluing horticultural work and other labour dismissed as unskilled requires a more holistic notion of skill which embraces all know-how, including that of nonhumans, and which recognises how political and economic forces shape work and learning in capitalist production spaces. Correcting devaluation of horticultural work requires we regard it as exceptional, not to excuse it from treating workers well, but to appreciate how essential its products – and therefore its workers – are to healthy people and planet, and how risky plant work is.

A focus on skill directly redresses misconceptions of horticultural work as unskilled, unravelling one foundation of its devaluing. It has practical value in helping understand the nature of the industry's skills crisis. Conceptually, skill helpfully connects analysis across the micro and macro: the bodily speed of the pickers' work bunching spring onions is required by the supply chain their employer feeds. Understanding the nature of skill in contemporary work might also point to how workers' lives could be enhanced.

Across these openings, I returned to how horticultural work has been imagined as unskilled. This is founded in flawed conceptions of knowledge and work, particularly in relation to engagements with plants. In contrast, I have set out an ecological interpretation of skill as socially embedded practice, enabling the achievement of a goal; horticultural know-how comprises the abilities required to grow food which extend across relations between people and others. Know-how developed on the job is central to production, meaning that lack of qualification or formal training cannot be a basis for regarding certain workers as unskilled, ignorant or inexpert. Horticultural skills are learnt and shared in practice, so countering skills shortages and ensuring sufficient access to food growing know-how comes through continuing to involve people in horticulture. Skilled horticulturalists cannot exist without the ability to practise horticulture. Horticultural know-how does not sit with an individual as it only emerges amongst interaction between grower, plant, colleagues and environment, including tools and materials. A healthy horticultural knowledge ecosystem therefore requires due attention to grower's and plant's whole situation, including economic dynamics. Given how notions of skill are deployed, particularly to categorise jobs and workers as unskilled, it is particularly important that skill is recognised as socially constructed by highly unequal power dynamics and practised within politically shaped contexts.

Outline of chapters

We begin at the scale of macro forces shaping growers' practices with a profile of the global horticultural workforce, its shortages and conditions. Chapter 2 explains how the labour regime dominating food production evolved with globalised agri-food supply chains driven by cheap food. Global fresh produce supply chains fed by industrialised production like that around the Mediterranean are only possible through cheap agricultural work. A food regime perspective shows how organisation of work and labour relations is determined by a production regime and capital accumulation which drive horticulture to minimise labour costs. This contributes to devaluation of horticultural work by disempowering producers and shaping precarious work conditions. These dynamics intersect with racialised and gendered inequalities because many horticultural workers are migrants who are socially and legally marginalised. I argue that whilst pay can be exploitative, it is conditions of precarity which encapsulate the injustice of reliance on migrant workers. Labour regimes predicated on temporary, flexible work have heavy social costs not reflected in the price of fresh produce or recognised by consumers.

Chapter 3 narrows onto UK horticulture, the shape of the industry and its workforce, showing it to be bifurcated by scale. I dive into a personal history of one horticultural region, and historic perspectives on agricultural work to trace the roots of its devaluation. We see horticultural work has typically been fulfilled by those marginalised in the labour force, particularly migrants following various routes into UK horticulture, including the current seasonal worker scheme. The chapter considers what the sector calls its skills crisis, meaning labour shortages and diminished knowledge infrastructure, and how this portrayal is shaped by social construction of the ideal horticultural worker. Counter narratives from parts of horticulture not struggling to attract new recruits suggest the crisis lies less with the nature of horticulture and more with the qualities of jobs offered. Across these debates, issues around skill are individualised, with individual workers and employers blamed, rather than looking to systemic causes requiring collective responses. The true crisis for UK horticulture lies within unjust and unequal agri-food systems, reliant on migration fuelled by global inequality and racial capitalism.

Chapter 4 delves into how workers are described by media and policy makers, analysing discourse around horticultural work and workers in the UK since the end of the 20th century. This period included considerable disruption from Brexit and the Covid-19 pandemic, which the industry portrayed as a series of unpredictable crises for which the solution is migrant workers. Again problems are individualised to avoid addressing structural inequalities. Looking to what horticultural work is really like, Chapter 5 details the skills required to grow crops, knowledge extending from close-up plant work to negotiating supply chains. We see wide-ranging skills and growers exhibiting care, judgement and dexterity. This includes fieldworkers with hands-on roles, who are portrayed in detail to counter suggestions that their work is unskilled. Growers sit at the meeting point of micro, meso and macro contexts as recognised by an ecological interpretation of skill and are never wholly in control of their work's outcomes because plants are unruly lives to work with. We see that some within horticulture regard it as unskilled because skill is perceived as an attribute of individual bodies. The counter view of skill as social practice is further developed through Chapter 6 which details how growers become skilled, considering processes of learning through doing and showing. Knowledge exchanges between peers and within workplaces are significant for skill development, but are pathways which can undermine the value placed on horticultural work. Good growers are never self-made, always reliant on more expert growers' skills, situating their practice within a social context. The power to grow food is a crucial condition for horticultural knowledge's flourishing, a power which can be highly constrained by wider circumstances.

This wider context is examined in Chapter 7 which compares the knowledge ecosystems archetypal to larger and smaller growers. It describes flows of knowledge around horticultural communities and where they are vulnerable or deficient, reducing resilience. But the skills challenges horticulture faces are not really knowledge issues, being rooted in globalised food systems which make it incredibly difficult to secure a good livelihood from growing good food. Across this chapter vignettes from specific growers connect archetypes to people's stories. In the final chapter, I look to further examples from UK horticulture, working towards commoning which moves us beyond the individualisation of risk and resources associated with horticultural knowledge. Those who grow food deliver common goods; this suggests shared responsibility to enable them to learn and grow. Horticulture's real crisis is the difficulty of making a viable income from producing good fruit and vegetables, and that so much of the produce we consume relies on the labour of undervalued, precarious workers. Imagining that the work they do is low or unskilled has helped perpetuate this devaluing; by the end of this book, I hope you recognise that horticultural work is deeply skilled and can help us understand the real nature of skill.

Notes

1 The Agricultural Wages Board for England and Wales was abolished in 2013 but Welsh Government established their own replacement; parallel bodies are in place in other devolved nations.
2 Project outputs are all available at https://www.cardiff.ac.uk/sustainable-places/research/projects/knowing-to-grow.

2 Global horticulture

Growing cheap food and precarious work

Introduction

The UN declared 2021 the International Year of Fruit and Vegetables (IYFV) in recognition of their critical contribution to human and environmental health (FAO 2021). It championed making horticultural produce universally accessible, as the foundation of a healthy diet and contributor to numerous Sustainable Development Goals (FAO 2021). Supply is currently inadequate to meet global health goals, but the year added pressure to increase production. Climate, health and environmental scientists recommend increased reliance on plant foods (Willett et al. 2019), and dietary transition also increases their consumption (FAO 2021, 40). Yet horticulture is vulnerable to changing climatic conditions, making maintaining current levels of production harder and the potential to increase supply uncertain. IYFV materials portrayed this demand as an opportunity, particularly for small producers who could switch from producing staples to profitable horticultural crops (FAO 2021, 23). But the desirability and feasibility of this switch are questionable given that fruits and vegetables require intensive management, knowledge and skills (FAO 2021, 23). Horticulture supplying local demand may enable rural development, but the dynamics around longer supply chains are likely to benefit corporations (Carolan 2011; Freidberg 2020). Global food systems do not typically profit those who grow fruits and vegetables.

The population of food growers is large: the global agricultural labour force was 866 million in 2021, just over 26% of total employment. This is despite the global trend of a declining agricultural workforce, with the sharpest drop (49% since 2000) in Europe (FAO 2022). Fruit and vegetable production remains labour intensive, so we can assume a significant proportion of this work is in horticulture. A picture of the European workforce is presented in Box 2.1. Horticulture generates a sixth of the value of USA farm sales but accounts for two-thirds of waged workers' pay, due to its labour intensity (Martin 2021, 21). Strawberry planting in California costs $30,000/acre and $20,000/acre to harvest, generating gross revenue of $60,000–70,000/acre, compared to $500–1,000/acre from Midwestern corn or soybeans (ibid).

DOI: 10.4324/9781003281689-2

Box 2.1 European horticultural work in numbers

Just 2.3% of farms in EU countries specialise in horticulture, but the distribution is highly uneven. Spain produces almost a quarter of the union's fresh vegetables, with Italy the only other member state producing a double-digit share. Other significant producers include the Netherlands, Germany and Poland.

Approximately 5% of EU workers are employed in agriculture. In 2022 this meant 7.8 million full-time agricultural workers, most of them non-salaried labour.

Foreign workers make up 24% of waged agricultural workers in Spain, 37% in Italy and 90% in Greece, not counting informal or illegal workers. As much as 61% of the EU agricultural workforce may be engaged informally. There are an estimated 150,000 undocumented migrants in Italian agriculture alone.

Horticulture and fruit production employ around 9% of the EU's full-time equivalent agricultural jobs on just 2% of agricultural land due to its labour intensity.

European farms specialising in horticulture generate almost three times higher average farm income, but by farm income per labour unit, they perform worse due to the intensity of waged labour.

Waged and seasonal workers are more numerous in horticulture than in other forms of agriculture. The figure of more than one third of all EU non-family labour working in the sector is likely an underestimate due to a lack of reporting from very small farms, seasonal roles and informal employment.

Workers in agriculture are more likely than average to have a lower level of education and to work long hours, with almost a quarter working 49+ hours per week.

(Figures from Carnibella and Wells 2022; Corrado et al. 2017a, 10; European Parliament 2019 & 2021; Eurostat 2022 & 2023; FAO 2022; Martin 2016).

As significant as the volume of jobs in horticulture is the nature of the work, as acknowledged in the FAO's introduction to IYFV:

> Jobs created in the sector should offer decent employment opportunities. But value chains for fresh produce are particularly vulnerable to environmental, social and governance risks. The industry has a relatively high proportion of informal workers (casual, migrant or family labour). It is not uncommon to find farms where workers toil long hours in difficult conditions, without adequate health-and-safety safeguards, and without

> respect for their rights, such as the freedom to form unions. Child labour has been frequently reported, gender equity is a problem, and gender-based violence is common.
>
> (FAO 2021, 43)

Despite noting characteristics the very opposite of decent work, nothing in the IYFV action plan targeted them, prompting an open letter to UN agencies, urging them to take a lead in promoting good work (Good Work for Good Food, 2021). The letter noted that images of happy smiling workers in IYFV materials misrepresent the realities of work in the industry for many people globally.

I highlight the IYFV discourse because it indicates a fundamental tension affecting horticulture: there are many good reasons for production to expand, which requires more labour, but horticulture already suffers labour issues. Many Global Minority countries were concerned with labour supply for food production before the Covid-19 pandemic caused a reputed shortfall of 1 million agricultural workers in Europe alone (FAO 2020b). Labour shortages are difficult to quantify because few statistics disaggregate horticultural work, and globally, most fruit and vegetable production occurs on small family farms with no waged workers (FAO 2021, 22). The sector's workforce is fluid with temporary and migrant workers often not captured by data, never mind undocumented and informal workers. We *do* know that many horticultural businesses in high-income countries struggle to attract the workers they need and are concerned about skills and labour shortages (CAHCR 2018; Horticulture New Zealand 2023; Pye and Tait 2020). At all career levels, across diverse roles, horticulture struggles to attract skilled people, and this limits business development (Aldous et al. 2014). Problems attracting labour are not peculiar to horticulture, as much agriculture in industrialised countries complains of severe shortages (Nettle 2018). It is difficult to specify the personnel shortages because of poor quality data and the lack of agreed approaches to define skills or labour shortages (Anderson et al. 2021). Labour and skills shortages are not the same thing (Nettle 2018), but are often conflated in public discourse (Pitt 2021b). It may also be in the industry's interest to cry shortage, as Campbell argues in the case of Australia where this narrative has been used to justify supplies of migrant workers, despite lack of evidence for the sector's skills needs (2019).

Data and forecasts from a country with significant commercial horticulture reveal a significant cause for concern. Canadian labour monitoring found that in 2017 horticulture employed 99,862 workers of which 43,483 were from overseas (CAHCR 2018). The sector reported 5,230 unfilled vacancies, calculated to result in CA$622 million lost sales for the year. Turnover rates were higher than average for agriculture, as were vacancy rates. Canadian horticulture's labour gap will widen, peaking at almost 60,000 by 2030, representing almost half the total national gap forecast for agriculture. This is in spite of projections that automation and efficiency gains will decrease labour intensity (CAHCR 2023). Employers suggested reasons for failure to recruit which emphasised the nature of the work – manual, seasonal, remote locations, outdoors – and employment conditions – relatively low pay, lack of career advancement – but also their sense of a generational shift away from interest in agriculture (CAHCR 2018).

The quality of jobs is significant here as it indicates why vacancies may be unfilled. European states with a legal minimum wage apply this to farm work, but historically, agriculture has been treated as a special case with reward structures typically providing lower incomes than other industries (Chapter 3). In many European states, agricultural workers are rarely represented by an independent trade union due to low rates of collective bargaining (Eurostat 2022), and where it exists, it often sets wages below the average for other sectors (Eurofound 2006). The greatest concentration of non-family labour is in fruit and citrus production which employed more than one third of the EU total in 2013 (European Parliament 2019). More than half of European citrus production is in Spain, locating this significant temporary workforce in the Mediterranean region which is synonymous with today's horticultural labour regime, as explained shortly.

The IYFV's failure to address tensions around expanding production despite worker shortages and exploitation was symptomatic of discourse around horticulture: fruits and vegetables are vital, but those who do the bulk of the work producing them are rarely visible, particularly when they are suffering. This paradox has been noted across food work which has always been the world's largest labour force, whilst often hidden (Böhm et al. 2020). Food and farm work has a long history of exploitation (Clutterbuck 2013). The horticultural workforce today is characterised by vulnerability in Australia (Underhill and Rimmer 2016; van den Broek et al. 2021), Canada (Barber 2014; Binford 2012), the USA (Gray 2014; Holmes 2013), the Mediterranean (Gertel and Sippel 2014) and across Europe (Rye and Scott 2018), including the UK (Rogaly 2008). Food workers are often invisible to consumers, including within movements promoting good food which neglected food workers, allowing consumers to imagine they were eating ethically (Allen 2016; Gray 2014). Blindness to conditions behind "local sustainable produce" was perpetuated by producers such as organic farmers who benefited from an imaginary of them as small family operations reliant on happy toil; meanwhile, they resisted controls which would improve conditions for their paid labourers (Guthman 2014; Minkoff-Zern 2017). Artisanal foods produced at small scales under agro-ecological principles have claimed superiority to conventional products from larger companies, even when offering waged workers no better pay or treatment, or relying on an unpaid workforce of trainees vulnerable to exploitation (Weiler 2022; Weiler et al. 2016). Critical scholars of alternative food movements, such as Julie Guthman, Patricia Allen and Margaret Gray, highlighted these contradictions. Meanwhile rural sociologists rediscovered a focus on agricultural labour, in particular immigrant workers in agricultural systems (Bonanno and Cavalcanti 2014; Scott and O'Reilly 2022), meaning we now have better insight into their situation.

This chapter outlines the structure of horticulture's workforce, placing European horticulture within its global context as it evolved since the late 20th century. European supply chains have significantly influenced UK horticulture and so are important context. The region's labour regime shares features with dominant global supply chains, but diverges from experiences and policies in the USA which have drawn the most attention in scholarship of food labour. Rather than providing a comprehensive history or global comparison, I aim to establish three fundamental characteristics of European horticulture and its workforce, key to understanding how this work became devalued: globalised supply chains which disempower producers and determine labour relations; reliance on migrant workers; and work conditions tending to precarity. These dynamics intersect with racialised and gendered inequalities, often resulting in worker exploitation.

Global supply chains and the secrets of cheap produce

Although language such as "global food" or "globalised agriculture" suggests otherwise, there is no single agri-food system or supply chain type (Allaire and Daviron 2019; Sage 2022). There is an equally complex landscape of terminology characterising food systems (Carolan 2011). What is clear is that how the world feeds itself currently is unsustainable and unjust, with productivism perpetuating the myth that the goal is to produce as much food as possible as cheaply as possible (Carolan 2013). In the corporate food regime (McMichael 2005), agricultural commodities are open to speculation and the dark arts of international financial trading. This focus on profit and accumulation most benefits corporations and investors (Clapp 2022; Zerbe 2019, 166). Outcomes are particularly poor for farmers who are under-rewarded by markets (Lang 2020). Of course, "cheap food" is a pure fairy tale, with the true costs paid in ecological impacts, climate crisis, poor human health and social inequality, particularly for people of colour. Cheap food can only be produced by cheap labour (Gerbeau and Avallone 2016, 128). And the cheapest labour grows fruit and vegetables, with over-exploitation of labour normalised in centres of intensive horticultural production (Gertel and Sippel 2014). Food system power has gradually shifted from producers (pre-1900), through processors (first half of the 20th century), to rest with retailers, with small-scale producers suffering particularly (Carolan 2011). Cheap food logics also make it hard to control labour standards, so workers suffer whilst corporations benefit (Lang 2020).

For horticulture, the outcomes are global supply chains with complex logistical arrangements and systems of actors supplying fresh produce to consumers year-round (Friedland 1994), including significant numbers of export oriented intensive producers which have expanded into low- and middle-income countries (Carnoy and Luschei 2008; Ngutu et al. 2018; Reigada and de Castro 2022; Selwyn 2014). Growing for global supply chains requires investment in technology, training and infrastructure less available to smaller farmers, so production is increasingly controlled by multinational corporations (Reigada and de Castro 2022). In the example of Brazil's dramatically expanded production of grapes for export, power gravitates to buyers, and highly capitalised producers are pushed to compete and to grow increasingly higher quality fruits to standards closely monitored by retailers (Selwyn 2014, 1023).

There are specific knowledge characteristics of this agri-food system. Corporatisation has extended into agricultural inputs, as companies acquire intellectual property rights over what were collectively held resources, such as seeds (Nally 2011; Zerbe 2019). Hands-off regulation preferred by neoliberal governments drove the rollback of state support for extension services and agricultural research which has become highly privatised as public investment declined (Anderson 2019). The IMF drastically reduced investment in R&D as innovation and expertise were privatised, concentrating control over key agricultural inputs and knowledge with a few transnational firms (Zerbe 2019). Investment in horticulture has suffered particularly as apparent through comparing the annual budget of an institute focused on just one cereal crop – the International Rice Research Institute, US$73M – with that of the World Vegetable Center, US$20M (FAO 2021, 31). There are no international research centres focused on fruit, and total capacity globally is dominated by commodity crops, like banana. The price-cost imperative of production in global supply chains is also a key determinant of skills ecosystems (Chapter 7), as fair profit does not reach producers, reducing their capacity to develop knowledge.

To understand how corporatised, globalised supply chains affect horticulture and its workers, I draw on characterisation of global food regimes. A regime perspective considers how the state-market relationship maintains capital accumulation through profitable agriculture (Corrado 2017). This approach is far from perfect, having under-emphasised trajectories outside the hegemonic regime and colonialism (Wilkinson and Goodman 2019). But food regime analysis began as an attempt to characterise changing patterns of production and consumption centred on the UK (Magnan 2012) and so places the country's horticultural industry in historic context. Comparing food regimes helps attend to shifting labour dynamics as work is structured by production regime and capital accumulation (Corrado 2017).

Grounded in political economy analysis of capitalism, food regimes give a comparative perspective on how agricultural production and supply chains are shaped by the hegemonic political regime (Gertel and Sippel 2014, 6; Magnan 2012). They began with Friedmann and Michael's (1989) account of two significant periods in agri-food history, explaining how agriculture interacts with the evolution of capitalism. The first international food regime (1870–1914) was characterised by trading between settler states and European imperial powers, with cheaply produced staples from Australasia and North America providing cheap food for workers in industrial centres. Food and agricultural workers moved internationally, or were moved in the case of enslaved people forced to provide labour in settler states, resulting in the international division of labour between coloniser and colonised (Magnan 2012). Friedmann and McMichael located the hegemonic power of this regime within the UK as it determined trade rules and was the largest food importer.

The global recession and world wars in the early 20th century created a crisis in the agri-food regime and undermined UK hegemony. Grain prices collapsed pushing national governments to intervene to guarantee prices and sustain farm incomes. This interventionist approach combined with industrialisation of agriculture, to produce the second mercantile-industrial food regime, and USA hegemony (Magnan 2012). This regime was characterised by the rapid modernisation of agriculture with state investment encouraging farmers to maximise production through intensification. As agriculture became reliant on machinery and other manufactured inputs, and more food was processed, food became cheaper and companies gained profit and influence. But this regime was unstable both in the face of the 1970s global economic crisis which undermined grain markets and because of inevitable consequences of its dynamics. Industrialisation drove farm consolidation because only larger, capital-rich farms could afford to invest in new technology. Greater reliance on purchased inputs reduced farmers' control over their production system and was often financed through debt, so expenditure grew faster than income (Deutsch 2012). Fewer farms meant a weaker farm lobby whilst agro-food corporations gained influence which they applied to resist regulation and secure trade liberalisation (Magnan 2012).

Views vary on whether we are now in a third food regime, and if so, its characteristics. Friedmann (2006) suggested the rise of a corporate-environmental regime, characterised by supermarket power to set private standards for food quality. Supermarkets both respond to and create consumer demand for foods differentiated by environmental or safety credentials, and quality, with food-focused social movements contributing to market segmentation (Magnan 2012). The late 20th century saw supermarket dominance ripple across Europe, bringing price wars and squeezing producers to sell cheaply whilst reducing their power to resist (Lang 2020). In contrast, McMichael (2005) describes the current regime as an evolved corporate one, characterised by liberalised global agriculture shaped by WTO rules. Others suggest the creep of financialisation into agriculture, including financial speculation of food commodities, defines capitalist relations shaping agri-food systems (Burch and Lawrence 2005). During this period, globalised horticulture has expanded into new countries as national governments in Global Majority countries seek to promote economic development through export production. Early phases of state support often benefit small farms, but typically evolve into encouragement for larger farms including those owned by multinational corporations (Selwyn 2010).

There is consensus that today's globalised agricultural production depends on a particular system of labour provision (Bonanno and Cavalcanti 2014; Reigada and Castro 2022). These agri-food systems arose with neoliberalism which saw the state step back from social intervention, and labour losing power to capital, resulting in cheaper, more controlled labour (Bonanno and Cavalcanti 2014). Labour relations are defined by flexibility for employers, creating vulnerability for workers, heightened by the mobility of most of the agricultural workforce. Reliance on migrant workers is presented as a market necessity, but is socially constructed to deliver workers preferred by employers, and enabled by immigration laws (Bonanno and Cavalcanti 2014). State and market continue the dance through which they seek to ensure agriculture generates capital.

As Corrado argued (Corrado et al. 2017a, b), each regime had particular labour dynamics drawing workers from peripheral regions to fuel production for core nations. The first regime depended on exploited rural workers in colonies and domestically, including large numbers of enslaved people. As agriculture intensified during the second regime, production depended on migrant workers recruited from former colonies to work in industrialised agriculture (Corrado 2017), a labour regime which has continued to consolidate. Today's agri-food labour is more thoroughly deterritorialised and heavily dependent on worker mobility (Corrado 2017). Global supply chains rely on waged workers in production and processing, and as demand outstrips local supply, the workforce has become mobile (Reigada and de Castro 2022). Employers need to flex numbers of workers and their hours in response to unpredictable just-in-time supply chains, hence flexible, temporary contracts (Gertel and Sippel 2014; Pollard 2012). Post-Fordist agri-food industries have also weakened trade unions through lower membership and neoliberal policies disempowering them (Cavalcanti & Bonanno 2014). The rise of global fresh produce supply chains was only possible through making agricultural work more flexible and precarious (Reigada and de Castro 2022). The agri-food labour regimes significant to horticulture have been characterised in relation to the geographic regions where they emerged and are prominent: the California model which evolved into the Mediterranean model.

The California model of horticultural work

Agriculture in the USA industrialised quickly and dramatically during the early 20th century, as the leadership class urged farmers to work like factories or businesses to maximise production (Fitzgerald 2003). Increasing efficiency drove farms to consolidate and specialise (Allaire and Daviron 2019). Scale and efficiency necessitated a larger workforce, so waged workers became the mainstay (Deutsch 2012). Production systems modelled on mechanised factories meant greater division of labour with complex jobs separated into simple tasks distributed across the workforce (Fitzgerald 2003). A specialist fresh produce industry blossomed in California, as improvements in transportation and refrigeration, plus innovations like the relatively sturdy iceberg lettuce, made it possible to sell fruits and vegetables further afield (Freidberg 2009). Growers became more capital intensive as they supplied ever longer supply chains – first into urban areas, later global exports (Friedland 1994). The "green gold" of California's high-value produce industry depended on cool chains reliant on refrigeration, making distributors more powerful and leading to their consolidation (Friedland 1994). It also needed cheap labour (Guthman 2014), particularly seasonal workers for intensive harvest work, who were mostly migrants from rural Mexico (Martin 2021). This hiring model has been associated with California since the 1950s when Lloyd Fisher characterised it as a structureless labour market determined by labour intensive crops and seasonal fluctuation in demand (Friedland et al. 1981). Although never unique to the state, California's fresh produce boom was particularly dependent on access to cheap, flexible labour (Freidberg 2009). Defining features of this regime include: impersonal relations between employer and employee; unskilled jobs widely accessible to the labour force; compensation by unit produced (piece rate) not time worked; limited mechanisation and low unionisation (Friedland et al. 1981). Smaller farms reliant on family labour could not compete with larger ones employing cheaper seasonal workers, so exited horticulture (Martin 2021).

During the Depression, a ready supply of labour came from farming families escaping the Dust Bowl; later the Bracero programme brought migrants from rural Mexico – 5 million between 1942 and 1964 (Martin 2021). By the mid-1970s, only around 20% of California's farmworkers had USA citizenship (Freidberg 2009, 121). The advantage for employers was that Bracero employees were tied to them and had fewer rights than citizens. The programme was closed to try and drive up wages for domestic workers, but producers instead changed production systems (Clemens et al. 2018). California's tomato growers mechanised harvesting and selected varieties to suit, driving consolidation in tomato growing and processing (Friedland et al. 1981). In contrast, lettuce production, which is particularly volatile and difficult to mechanise, drew on casual and illegal employment as people remained cheaper than machines. Legal immigration from Mexico peaked in 2000, whilst illegal immigration declined after the 2008 recession, meaning labour costs increased. But cutting off immigration has not necessarily increased wages, as growers construct labour shortages to lobby for a guest worker programme, or take fields out of production. The industry also responded by increasing imports from countries where labour is cheaper, including Mexico (Martin 2021).

One feature of the California model is terrible conditions for horticultural workers. Latinx workers on strawberry farms were found to experience multiple health risks as fieldwork requires constant stooping which is physically demanding and harms bodies (Holmes 2013). Workers earn piece rates and so seek to go as fast as possible and avoid breaks, compounding physical strains. Pickers complain of racial abuse by supervisors, or having work under-counted to lower their pay, but as contracted labour it is hard for them to resist (Holmes 2013). Farm unions have periodically mobilised farmworkers to secure improved pay and conditions, most notably with the National Farmworkers Association (now United Farm Workers) under Cesar Chavez in the 1960s. But Martin (2021) suggests it has been difficult for collective action to secure workers' rights under the California model. Weapons typically available to unions, such as strikes, are less effective because of inelastic demand for farm commodities, plus the availability of undocumented workers reduces the legal workforce's leverage. USA unions are quite top down and so lack direct contact with farmworkers, particularly temporary crews employed via intermediaries. Federal labour laws lack provision for unionisation in agriculture, due to historic assumptions that it relies on family labour not requiring legal protections (Martin 2021). Farm owners – including organic – have also sought to maintain access to surplus labour to reduce workers' influence and resisted tighter labour standards (Guthman 2014, 2017).

Today's California model is multi-layered, with varied hiring practices and actors often obscuring workers from farm owners (Martin 2021). Guest workers on seasonal visas mix with unauthorised and legal workers, including settled migrants. Guthman suggested the Bracero programme institutionalised low wages in Californian horticulture and habituated growers to an oversupply of cheap, disposable workers (2014). But she notes that growers are not wholly to blame as immigration policy provided the highly skilled, cheap, docile workers they wanted, plus growers need to minimise labour costs to survive. These are systemic problems:

> The corporatization of U.S. agriculture and the growth of international free markets squeeze growers such that they cannot easily imagine increasing the pay of the pickers or improving the labor camps without bankrupting the farm. In other words, many of the most powerful inputs into the suffering of farmworkers are structural, not willed by individual agents.
>
> (Holmes 2013, 52)

An agri-food regime driven to produce food cheaply produces labour relations like the California model, with working conditions and wages subordinated to profit maximisation and price minimisation (Gerbeau and Avallone 2016), disempowering workers in favour of corporations (Allaire and Daviron 2019). As growers struggle to remain profitable, labour is the input where they can make savings, hence the pursuit of ever cheaper, more flexible labour. The same dynamic enabled the Mediterranean region to become a global centre for horticulture, with employment regimes intensifying the worst tendencies of the California model.

The Mediterranean model

The world's largest concentration of greenhouses in Almería, Spain is famously visible from space, a dubious achievement made possible by carrying "the intensive agricultural model to its extreme in terms of its density, economic scale, environmental side effects, and social implications" (Gertel and Sippel 2014, 17). By 2012, 29,000 ha of greenhouse – a dramatic increase from 1975's 3,000 ha – made horticulture a key economic contributor for the region. Almería produces around 3 million tonnes of vegetables annually, largely for export within the EU (Aznar Sanchez et al. 2014, 112). It has very low water availability and is distant from north European markets, but has other locational advantages: eligibility for EU structural funds and access to migrant workers. The latter is significant as each hectare of production requires on average 560 working days annually, representing 2.3 FTE workers, and around 40% of costs (Aznar Sanchez et al. 2014, 118). These conditions allowed dramatic expansion of capital-intensive horticulture which Spanish governments encouraged to promote economic development by investing in knowledge extension and infrastructure (Veiga 2014).

Symptomatic of the corporate food regime, supermarkets gained power across the latter part of the 20th century, growing in scale and market share (Burch and Lawrence 2005). Their influence on fresh produce has been particularly marked as they dictate what is produced and quality standards, using low prices to attract customers (Gertel and Sippel 2014). Small family farms which had historically featured across the Mediterranean could not compete and so production consolidated (Gertel and Sippel 2014; Sippel and Gertel 2014). The complexity and technological requirements of supplying supermarkets expanded the role of intermediary actors such as wholesalers and shippers, shifting power and profit from producers. Production therefore intensified and specialised, meaning fewer bigger farmers, with workforces dramatically shifting from reliance on family labour to migrant workers (Sippel and Gertel 2014; Veiga 2014). Some medium-scale producers remained competitive, but only through over-exploitation of cheap migrant labour (Corrado et al. 2017a).

Production around the Mediterranean now dominates European horticulture, meaning its labour regime prevails, not only there, but through pressure on producers elsewhere to compete and therefore to reduce labour costs. Like the California model, this labour force is characterised by fluid employment practices which prevent close, stable ties between workers and employers (Veiga 2014). One or two sub-contractors positioned between worker and boss, cloud responsibility for ensuring fair treatment, and reduce workers' ability to push for improvement. Most of the workforce is temporary, a factor which combines with oversupply to keep wages low and control high (Sippel and Gertel 2014). National governments established migration pathways for seasonal workers to ensure supply, the first wave mostly from Morocco and Eastern Europe, followed by those from other parts of Africa. Migrant workers are required to return home after harvest and are deliberately located away from the settled population to reinforce the sense that they are not full citizens (Veiga

2014). Visas often tie the worker to a particular employer, compelling them to work even when treated badly (Sippel and Gertel 2014).

Greenhouse work in Mediterranean countries entails harsh conditions including exposure to harmful chemicals or long hours on tasks requiring physical exertion and stamina. Workers who resist have faced violence: in 2013 Bangladeshi strawberry pickers in Greece demanded six months' worth of pay withheld by their employer; 33 were shot by their supervisor with several seriously injured (Papadopoulos and Fratsea 2018, 140). In other areas tensions between workers or between migrants and locals have erupted into violence; workers have also died because of work conditions (Corrado et al. 2017b). But Mediterranean horticulture depends on "an ever cheaper and more flexible workforce" which means foreign migrants (Sippel and Gertel 2014, 47). As with their Californian counterparts, growers would argue they cannot increase labour costs and remain economically competitive: labourers' insecurity is entangled with that of producers. Agriculture's dependence on international labour mobility was also enabled by immigration policies (Corrado et al. 2017b), so capital and state collaborate to exploit agricultural workers (Rogaly 2021). It remains to be seen whether new EU legislation intended to re-balance supply chain dynamics will alter this to benefit workers.

Horticulture's reliance on migrant workers

The previous sections indicated that two of the world's most significant horticultural zones rely on migrant workers, with production systems and economic viability dependent on supplies of cheap, mobile, flexible workers. This recurs globally (Box 2.2). The fates of migrant workers and agriculture are so closely entwined that Agyeman and Giacalone propose an immigration-food nexus (2020). It is only possible to produce year-round crops of strawberries and tomatoes in vast polytunnels around the Mediterranean because inflows of workers from North Africa fulfil low-paid seasonal roles (Corrado et al. 2017a). Reliance on migrant labour (Figure 2.1) is the latest phase of global agri-food

Box 2.2 Migrant food workers globally

Europe

An estimated 800,000–1 million seasonal workers work in European agriculture annually.

Germany receives around 300,000 workers annually for agricultural, horticultural and forestry work, mostly from within the EU.

The UK employs around 75,000 seasonal migrants for horticultural work. Seasonal worker visas are now available to EU and non-EU citizens.

Sweden's berry industry receives 3–5,000 seasonal workers on temporary visas, mostly from Thailand.

In Italy, 400–500,000 migrants from 155 countries are employed in agriculture. An estimated 40% are irregular or illegal workers, although in some regions this is thought to reach 90%.

France and Spain employ 276,000 and 150,000 seasonal agricultural workers, respectively. A large share of visas go to North African states through bilateral agreements.

Beyond Europe

In 2018, 75% of USA farmworkers were migrants. Temporary migrant workers are estimated to contribute 10% of crop farmworkers. Estimates of undocumented farmworkers in the USA range from 1.5 to 2 million, or 50–70% of all farmworkers.

Canadian horticulture employed 43,483 foreign workers in 2017, 43.5% of its workforce, a proportion which has steadily risen since 2005.

In Australia, an estimated 50% of labour in vegetable farming and 30% in fruit and nut farming are temporary migrant workers.

(Figures from: Agyeman and Giacalone 2020, 6; CAHCR 2018; European Parliament 2021; FAO 2020b; ILIO 2016; Mieres and McGrath 2021)

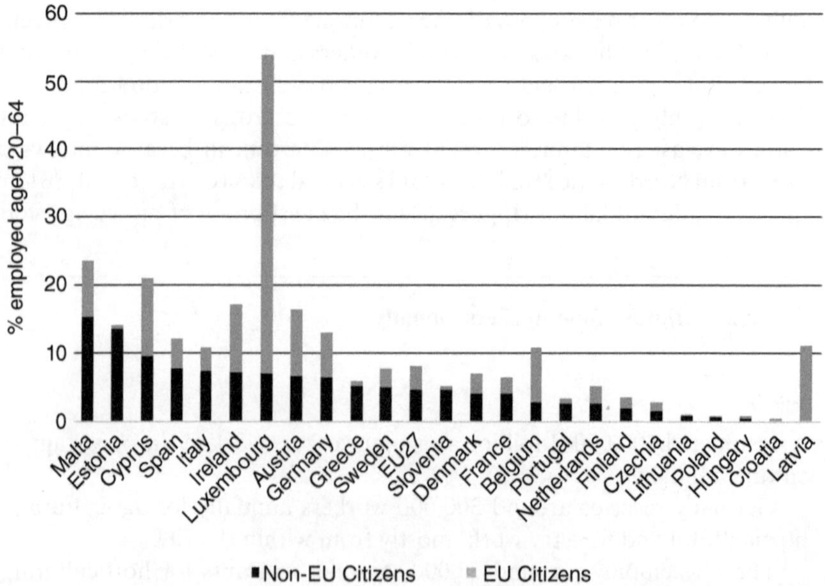

Figure 2.1 Composition of the EU's agricultural workforce in 2020 (Eurostat 2020).

systems' reliance on spatial fixes to supply labour, with the evolving dynamics of colonial exploitation and slavery (Corrado 2017; Gerbeau and Avallone 2016; Gertel and Sippel 2014; Scott 2022). Export-driven production in Global Majority countries uses their cheap labour to feed consumers in higher income countries (Fischer and Benson 2006), a dynamic some compare to plantations (Martin 2021). The inequalities of racial capital (Gilmore 1999), colonial racialisation and capture of labour remain endemic (Rogaly 2021).

Food growing globally employs considerable numbers of migrants, with varying organisation and conditions (Martin 2021). A recurrent feature are guest worker programmes offering temporary immigration or targeted visas linked to seasonality of labour intensity (Table 2.1). These were first used in 19th century Prussia and South Africa, then spread through other nations concerned to balance the need for cheap workers with anti-immigration sentiments (Hahamovich 2003). The USA Bracero programme commenced in 1942, employing mostly Mexican men, and was followed by post-War schemes in Europe which mixed reconstruction, cultural exchange and refugee employment. Canada's Seasonal Agricultural Worker Program (SAWP) is one of the most long-standing, employing Latinx and Caribbean people via highly controlled bilateral state agreements since 1966 (Binford 2012). Often regarded as the global standard in protecting seasonal workers, SAWP still offers no path for migrants to become citizens (Martin 2021) and does not fully protect workers from exploitation or unhealthy conditions (Binford 2012; Weiler et al. 2017). Rapid growth in New Zealand horticulture this century relied on seasonal workers, mostly from Pacific nations (Gibson and Bailey 2021). Similar routes were established slightly later in Australia, where a greater proportion of work is done by backpackers (van den Broek et al. 2021), as the working holiday visa incentivises time working in horticulture (Walsh 2014). Until recently USA horticulture was notable for relying on undocumented Latinx migrants, but since the turn of the century, levels of unauthorised work have declined due to the recession and tighter immigration controls. In 2019 legislation was introduced to regularise seasonal workers and offer the potential to move from a seasonal visa to a longer work permit (Martin 2021).

Northern European countries also have seasonal migrant agricultural work programmes (Scott and Visser 2022). Martin suggests the region has some of the world's strongest labour regulation and social security, meaning fewer "factories in the fields" than in the USA (2021). The EU's Seasonal Workers Directive established equal treatment with domestic citizens in relation to employment and social security benefits linked to sickness, but not those for unemployment or family. However, enforcement varies, and inspections often find violations (European Parliament 2021). Well-regulated guest worker programmes and employment of intra-EU migrants across Northern Europe have in their shadow Southern Europe's reliance on non-EU migrants with less regular conditions and widespread exploitation. For example, the Spanish strawberry industry employs large numbers of women from Morocco via a

Table 2.1 Comparing seasonal worker programmes

	Canada	*New Zealand*	*Spain*	*USA*	*UK*
Scheme	Temporary Foreign Worker Programme – Primary Agriculture	Recognised Seasonal Employer Scheme	Various, including Temporary and Circular Labour Migration bilateral agreements	H-2A Temporary Agricultural Workers	Seasonal Worker Visa (Horticulture)
Cap on numbers	71,000 (2022)	19,000 (2022)	No cap. Approx. 20,000 visas issued in 2014	No cap. Approx. 300,000 visas issued in 2022	45,000 (2022)
Recruitment	Sending government (Latin America, Caribbean) prepares lists, employer selects, 80% returnees	Employer direct or via sending government lists, targets Pacific Islands	Via agency recruiting from target countries	Employer direct from 80+ countries; must show failure to recruit USA citizens	Via authorised operators; no specified countries
Travel costs	Shared	Employer pays half of airfare	Employers pay and then deduct from wages	Employer pays all	Employee pays all
Visa duration	Up to 8 months; tied to specific job	Up to 7 months	Up to 9 months	Up to 3 years	Up to 6 months
Employment conditions	Housing provided; guaranteed national minimum wage and minimum work hours	Housing provided but employee pays; guaranteed national minimum wage and minimum work hours	Guaranteed 18 6.5 hr days work monthly at minimum wage (€34/day)	Tied to job; housing provided or pay housing allowance; minimum wage requirements vary by state and size of farm; average pay US$13/hr	Specific job only; earn a national living wage; accommodation often provided and charged for by the employer
Issues	From 2020 seasonal workers have a pathway to permanent residence	Cited by ILO as good practice	Workers complain employers over-recruit and use withdrawal of work time to discipline	Lawsuits suggest employers do not abide by visa conditions; less productive workers are likely to be fired; dependents can accompany workers	Increasing reports of malpractice and worker exploitation

Adapted from Martin (2016) and DEFRA (2023).

Sources: CAHCR 2023; DEFRA 2023; Horticulture New Zealand 2023; USDA Economic Research Service 2023.

bilateral agreement. Workers' children cannot enter Spain, and workers must return home at the end of the season. Producers deliberately over-recruit so workload is insufficient to earn a good income, and those tempted to resist poor treatment – including sexual harassment – are discouraged, fearing work will be offered to more pliant women (Hellio 2014). In Southern Italy, these dynamics are heightened through greater use of undocumented workers and criminal gangs' involvement in hiring (Carnibella and Wells 2022; Corrado and Palumbo 2022; European Parliament 2021).

Globally, seasonal worker schemes established as a supposedly emergency solution to acute labour shortages have become systemic, being growers' preferred way to recruit (Chapter 4). The schemes became popular because they meet multiple priorities and create the "perfect immigrant" (Hahamovich 2003, 72–73). Businesses get workers who are bound to them and can be disciplined by the threat of deportation; trade unions are placated by tying migrants to onerous work; government gets immigrants who do the work and then go home; host countries get labour without the burden of the person or their family; and workers get income which benefits their home economy. However, they do not acquire citizenship, and their complete needs as persons, such as the right to family life, are not part of their visa. Many live in on-farm accommodation separate from the local community, who may prefer them to be un-integrated, whilst employers can use spatial segregation to keep workers unaware of their rights (Scott and Visser 2022). Recruitment is often via intermediary agents who make it difficult for workers to appeal for better conditions or unionisation (Caruso 2017; Martin 2021). Treatment in many European countries is relatively good as states compete to attract workers (Hahmovich 2003). But the workers remain temporary or partial citizens, lacking the freedom to move between employers or into other employment, whilst living under threat of deportation, all of which is state sanctioned (Binford 2012; Hahamovich 2003; Stead 2021; van den Broek et al. 2021; Weiler et al. 2017).

Conditions for horticultural workers are well-documented, leaving no doubt that a labour model reliant on under-valuing highly replaceable workers of colour feeds European supermarket shoppers (Corrado et al. 2017b; Gertel and Sippel 2014). Take tomatoes grown in greenhouses across Southern Italy and Sicily, a popular product across northern Europe. This vast industry constantly faces pressure from processors to reduce prices, with the "double down" auction system setting a price before production begins (Melossi 2021). Traditional labour contracting via *caporalato* evolved from a system recruiting day labourers from surrounding districts, to manage migrant workers, mostly from sub-Saharan Africa (Melossi 2021). The *caporale* acts as an intermediary between workers and growers, taking a proportion of the team's pay (Avallone 2017). *Caporali* recruits migrant workers subject to intense workplace control, earning on average €20–30 daily for long hours in inhuman conditions. Tomato workers live in shanty style housing regarded as ghettoes, conditions witnessed in 2020 by a UN rapporteur:

> In my visit to a ghetto in Apulia, I witnessed the terrible working and living conditions for agricultural migrant workers where some lived in overcrowded accommodation with scarce access to water and without access to electricity, sanitation or basic needs. The ones arriving for the harvest season would have to live in tents or hand-made shacks. The number of irregular agricultural seasonal workers would double during the harvest season making conditions even more difficult. Some of them stated that without the support of organizations such as CARITAS they would not be able to survive.
>
> (OHCHR 2020)

Pay is often below the minimum wage, with some illegally paid piece rates. Workers are exploited through racialised discrimination, portrayed as criminals in the media and subject to disproportionate, sometimes violent policing (Melossi 2021). Whilst essential to the tomato industry, they are not fully included within local society as they lack citizenship rights and are subject to deportation if found to lack immigration papers. Their "outsider" status is spatialised in shanty towns located beyond the settled population who regard them as problematic migrants rather than valuable workers. Workers may depend on a *caporale* for essential documentation or access to housing, further reducing their power (Avallone 2017). Precarity creates competition between workers as each fears another will take their work, creating tension which accelerates work pace, resulting in faster, more efficient production which employers want (Piro and Sanò 2017). Precarious workers are good workers – they work harder and are more compliant.

The precarity of horticultural work

> People who will not eat it, and cannot afford to purchase it, under conditions of very limited political and social power, harvest much of the world's food supply.
>
> (Deutsch 2012, 74)

Food justice scholars and movements now recognise how food systems reliant on migrant workers are unjust, for dependence on marginalised groups' labour, and failing to protect workers from food insecurity. The preponderance of bad jobs within food production is a feature of global agri-food systems laden with racist, patriarchal, capitalist injustices (Klassen et al. 2023). Migrant workers do a huge proportion of the work across global food systems because it suits employers to recruit marginalised people, and those without permanent citizenship or legal immigration status are particularly marginal (Agyeman and Giacalone 2020). Migrant food workers are marginalised spatially as their employment often takes them to remote locations, and socially as they have precarious employment and struggle to integrate, or suffer racism (Lever and Milbourne 2017). For horticulture, the injustice of workers' treatment is embedded within production systems as industrialised production requires seasonal employment which offers no security (Gertel and Sippel 2014). Intensive horticulture around the Mediterranean only became possible through access to migrant workers who represent a neoliberal labour regime par excellence.

Horticulture's reliance on migrant workers is unjust in two interlinked ways. Firstly, international migration pathways exist and continue because of global inequality which keeps some countries poorer than others, making it desirable for migrants to move towards higher earning potential. People migrate to agricultural work in the hope of better lives because they are on the wrong side of global inequities (McLaughlin and Weiler 2017). Labour mobility only works as a spatial fix for production in high-income countries because other regions and nations are economically poorer (Scott 2013). Secondly, their status as immigrants disempowers workers and leaves them vulnerable to exploitation, particularly when their immigration status is temporary and tied to agricultural work. As Constance articulates, most important about the agri-food system's reliance on immigrant labour are:

> new dimensions of insecurity and vulnerability [are] associated with this labour regime based on flexible accumulation by market actors and neoliberal restructuring by state actors. The renegotiated state-market nexus creates hazardous spaces which labour occupies.
>
> (2014, 163)

These labour regimes give employers control over workers without requiring responsibility or liability in return (Constance 2014), making horticulture a nadir of neoliberal labour relations which tend to precarity.

Horticultural workers do not always earn a fair wage (Box 2.3), symptomatic of industrial agriculture which offers food workers "minimally decent compensation" (Barnhill and Doggett 2018, 8). Capitalism depends on cheap food, which requires industrial food regimes dependent on cheap labour. Corporate consolidation of power over agri-food supply chains has heightened this by reducing producers' ability to set prices, reducing their income and driving them to reduce costs (Clapp 2022). But another feature of this workforce is perhaps a more significant distinguisher of the current labour regime. Just-in-time supply chains in which retailers dictate conditions and schedules further loosened the ground under the food workforce, shaping labour regimes predicated on insecurity for workers and flexibility for employers. Supply chains pressure horticulture particularly, leaving their workers most vulnerable.

Horticultural markets are buyer-driven, making producers vulnerable to sudden changes in buying practices or prices, reducing their bargaining power (Schenner 2018). Fruit and vegetables have limited shelf-life, and supermarkets want orders delivered just-in-time, further intensifying pressure on profit margins. Labour inputs are where producers have some flexibility to drive savings (Hingley 2005; Rogaly 2008; Scott and Rye 2021; Schenner 2018), and labour

Box 2.3 How much do horticultural workers earn?

The average hourly wage for a USA seasonal agricultural worker is around US$13/hr.

UK workers are entitled to national living wage, £10.42/hr. Median pay for workers in horticulture was £19,700 in 2019, below that of farmers (£26,200), although managers earn on average £32,400.

Nordic countries have union-agreed rates for horticultural work, but workers do not always receive these, either through employer deception or workers' preference for piece rates.

Italy has union-negotiated rates for agriculture of €42–48/6.5 hr of work, but workers often receive less. The UN Special Rapporteur on Right to Food found harvest workers earning €5–50 for a working day averaging 6.5 hours. Outside harvest season, they might earn €2–3/hr.

Workers may be offered piece rates per unit of performance, such as weight of picked produce. In the UK, these can only supplement legal wage rates, meaning productive workers earn more. But workers report not being paid for everything they pick, or being disciplined for working slower than the rate covering the cost of minimum wage by having their hours reduced.

(Figures from: Carnibella and Wells 2022; DEFRA 2023; Kuns et al. 2023; MAC 2020; McAndrew et al. 2023; Martin 2016; OHCHR 2020)

costs can be minimised by using temporary workers or flexible hiring, avoiding paying workers for any time they are not generating value (Rogaly 2008, Rye and Scott 2018). The profit squeeze pressures growers to seek the most productive, cost-efficient labour which tends to be migrants (Rogaly 2008), who are "reliable, flexible and compliant" (Milbourne and Coulson 2021, 500). The result is a just-in-time labour regime – the flexible and impermanent workforce of highly fragmented jobs, with informal or variable conditions, symbolised by an agency worker on a zero-hours contract (Bonanno and Cavalcanti 2014; Gertel and Sippel 2014; Pollard 2012; Reigada and de Castro 2022). Workers are unlikely to have collective bargaining rights, as unions have been deliberately undermined (Cavalcanti and Bonanno 2014). This keeps wages low, as does segmentation of the workforce along lines of migration status; high levels of migrant work are caused by and reinforce employer flexibility equating worker precarity.

Cavalcanti and Bonanno (2014) outline multiple dimensions of the flexibility so important to horticultural employers. Flexible work time entails workers' discontinuous employment and requirement to work variable hours, responding to demands for work or rest at short notice. Workers' activities and conditions are flexible as they are moved between tasks and lack the right to guaranteed hours or pay. The mobility many agricultural workers practice as they move between work and home is another form of flexibility. These combine to shape heightened flexibility for the employer which creates precarity and vulnerability for workers (Mésini 2014; van den Broek et al. 2021). Gertel and Sippel coin the term *flexicurity* to highlight this combination of flexibility and insecurity for workers (2014a, 11). Flexible workers are often recruited through indirect hiring by intermediary agents. Contracting teams for specific periods of work gives growers flexibility they need to minimise labour costs (Boeckler and Berndt 2014, 30). Intermediaries such as Italy's *caporali* enable flexible recruitment, whilst keeping workers in a dependent position where they lack the freedom to find better work (Avallone 2017). The Swedish berry industry grew its export market through switching to labour provided by recruitment agents with complex brokering chains reaching into rural Thailand. Each actor takes a portion of the labour value, reducing pickers' income (Hedberg 2014).

Reigada and de Castro (2022) regard diverse recruitment channels, combining agency contracting with undocumented workers and seasonal worker programmes, as a key feature of the fresh produce labour regime, providing flexibility for growers to hire for short periods, whilst encouraging workers to work longer to increase their earnings. Contractors or agents sever connections between employer and worker, reducing the workers' ability to push for better conditions (Cavalcanti and Bonanno 2014). Flexible hiring favours those who employers regard as having demonstrated satisfactory performance, which means being controllable and productive (Pedreño et al. 2014). The resulting intensification plays out as a spectrum of exploitation in which most employers operate within the law, but minor abuses present a slippery slope into more extreme infractions including abuse and forced labour (Davies 2020; Schenner 2018; Reigada and de Castro 2022). At their worst, this includes child labour and conditions of modern slavery (Clapp 2022). The ILO estimates that 71% of the world's 152 million child labourers are illegally employed in agriculture (Martin 2021, 13). With so many workers being migrants, it proves easier to get away with abuse (Milbourne and Coulson 2021).

Workers' vulnerable position, I would argue, is more significant than – although entangled with – low pay as a signal of horticultural workers' devalued status. Earnings reflect agriculture's tendency to offer lower than average incomes, but in most European countries minimum wage legislation should guarantee horticultural workers' pay. In the UK, pay is relatively well regulated, and good seasonal workers earn well (Chapter 4). Of course, not all employers pay legally; some reduce workers' take-home pay through unfairly high accommodation charges. Piece rates can increase workers' earnings, but also favour employers by incentivising fast, long work, and can be manipulated by withdrawing shifts to discipline workers. The UK's situation is parallel to Dutch horticulture's labour regime which Siegmann et al. (2022) term regulated precarity as the insecurities workers face are shaped by state regulation. Migrant precarity is legalised in practices including zero-hours contracts, wage discrimination, disciplinary fines, unhealthy work conditions and limited opportunities for personal development or progression.

Kuns et al. (2023) argue that agriculture's temporary workers are the quintessence of precarious work, meaning a lack of job security, limited economic and social benefits and a lack of legal protections. Employers seek marginal groups to "maintain a docile and highly controllable labour force" (Bonanno and Cavalcanti 2014, xxiii). During the 1990s, Moroccan agricultural workers in Spain were given legal status which enabled them to demand better conditions; as a result, employers looked elsewhere for more submissive labour (Gadea et al. 2017). Control is aided by migrant worker programmes which limit workers' ability to switch employers in pursuit of better treatment, and employers can threaten the loss of work or dismissal as disciplinary measures. When accommodation and right of residence are tied to the job, workers are essentially engaged in unfree labour relations, endorsed by governments which set the terms of seasonal worker visas (Preibisch 2012). Employers can recruit for such roles because neoliberal employment laws allow exploitation of food workers (Rogaly 2021; Strauss 2013). In 2011 the EU legislation to protect migrant workers brought their rights in line with those of naturalised citizens, but excluded seasonal workers (Mésini 2014). State protection is stronger in countries like the UK and Canada with fewer undocumented workers and stronger labour laws (Martin 2021). But UK consumers should not assume their produce was produced through fair work (Chapter 3).

Labour regimes predicated on temporary, flexible work delivered mostly by migrant workers create heavy social costs not reflected in price or recognised by consumers. Labour practices across Mediterranean horticulture depend on the unpaid labour of social reproduction, including caring for children whilst workers focus on waged work (Gertel and Sippel 2014). The social costs of fresh produce include physical and emotional strains on workers' bodies (Gertel and Sippel 2014; Stead 2021). From 2025, all European farms which receive CAP payments will be subject to social conditionality measures intended to protect workers' rights (Directive (EU) 2019/1152). Employers will be required to communicate to workers, including those on flexible or open contracts, their likely hours, conditions of work and rights; they must ensure workers are safe and given due health protection. It remains to be seen if these requirements will reduce seasonal migrant workers' precarity, particularly as they imply a duty to *inform* workers of flexible arrangements rather than curtailing them. Plus, employers who currently abuse and discriminate against migrants already break the law. Moves to control irregular employment practices within the EU might drive production to states with fewer labour controls; production in Morocco has already increased because of lower labour costs (Sippel 2014). Many governments are keen to welcome fresh produce companies bringing economic outputs and jobs, and typically favour the businesses' interests over those of workers (Martin 2021).

Horticultural workers at the intersections of injustice

So far this chapter has shown how global fresh produce supply chains depend on migrant workers, employed via flexible mechanisms favouring employers, creating precarity for workers. As examples from the Mediterranean indicate, horticultural workers often occupy positions at the intersection of race, gender and migration status, intensifying their marginal status, making them more controllable and vulnerable to exploitation (Barnhill and Doggett 2018, 8). Women are often excluded from better paid jobs, whilst undocumented workers are disciplined through threats of criminalisation (Reigada and de Castro 2022). Advantages such as permanent contracts are reserved for domestic workers (Avallone 2017), and roles are allocated according to highly gendered and racialised assumptions about who suits which task. The 1930s California lettuce industry employed "stoops", so called because short-handled hoes used to tend the crop required workers to continuously bend over. Growers argued that Mexican, Filipino and Japanese immigrants physically suited this role, so higher pay would not attract local white workers to replace them. In contrast, better paid, unionised roles in packhouses were dominated by white Americans; when the depression hit, the wages of women and migrants were reduced (Freidberg 2009). Ethnic segregation continues in Mediterranean horticulture: producers in Almería justify paying Romanians more than Moroccans on the ground that they work harder. Through such racialised stereotypes, employers keep the workforce segregated to foster competition between ethnic groups (Pumares and Jolivet 2014), reducing their collective power (Reigada 2017).

In many horticultural centres, the workforce has become feminised, with women dominating produce handling roles. Employers justify this through portraying such tasks as requiring small hands and a delicate touch, but women's real advantage to them is the potential to pay lower wages and that they are less attached to the labour market (Bonanno and Cavalcanti 2014, xxxi). Since the early 2000s, Moroccan women have been central to strawberry farming in Andalusia, via a bilateral state agreement which targeted married women with young children, to provide confidence that workers would return home after the harvest (Hellio 2014). This recruitment is also gendered in employers' expressed preferences for women's more delicate bodies which better handle soft fruit, and for supposedly feminine traits like patience and physical endurance which enable them to work long hours on repetitive tasks. Employers' justifications for their feminised workforce reflect expectations that Moroccan women are docile and pliant, meaning they accept poor conditions including low pay and abuse (Nieto 2014). To justify low pay, employers deploy stereotypes that women's extra-domestic work is supplementary income not essential to the household (Pedreño et al. 2014). Flexible employment via short-term contracts requiring women to return home at the end of the season is justified through sexist constructions which present flexibility as desirable for them as they want time for family duties, but the real benefits are women's precarity and undervalued contribution to profit (Cavalcanti and Bonanno 2014).

Treatment of marginalised horticultural workers implicates fresh produce in unjust global agri-food systems underpinned by interlocking inequalities of capitalism, colonialism and racism (Cadieux and Slocum 2016; Sbicca 2018). Just food systems cannot rely on production which exploits producers and workers likely to be people of colour, and which least protects people with intersecting marginalised identities (Garth and Reese 2020; Saville and Adams 2021). An increasing number of academics and scholar-activists pursue food systems more just for essential workers (Klassen et al. 2023). Achieving good food work requires the correction of global structural inequalities extending beyond food systems. This is hard to fight for when food workers are easily overlooked by consumers because capitalism so effectively hides them (Gerbeau and Avallone 2016). Retailers are not ignorant of risks of labour exploitation in their supply chains, hence the rise of ethical codes and audits supposed to control it (Freidberg 2003). But voluntary certification is more about reputational management than protecting workers (Brown and Getz 2008), with retailers not paying to guarantee fair conditions (Freidberg 2003). Complex food systems in which produce and workers are traded between intermediaries make it particularly difficult to know where food has come from, whose hands tended it or whether they were safe from harm. Farms with particularly pernicious work conditions are not isolated examples of the system failing, rather the system doing what it was designed to. But when such cases gain attention, they are called bad apples rather than blaming the whole rotten barrel, because cheap apples are good apples.

Workers themselves are acutely aware of their treatment and its injustice, hence worker-led movements fighting for change (Alkon and Agyeman 2011). But collective organising has been limited by agricultural employers' lobbying for exceptions from labour regulation. Precarity also makes it difficult for workers to resist and unionise (Constance 2014), so farmworkers are disempowered relative to employers (Brown and Getz 2008). Movements such as the Coalition of Immokalee Workers (Gottlieb and Joshi 2013) and those inspired by it in Europe (Caruso 2017) have combined their resistance with consumer power to pressure retailers to force producers to treat workers better. Such progress is heartening, but struggles in the face of the systemic drivers of agri-food labour regimes which disempower producers and undervalue workers.

Conclusion: The injustices of horticulture's undervalued 'unskilled' workforce

This chapter has shown how the current agri-food regime dominated by corporate retail power fed by global supply chains disempowers producers, pushing them to squeeze labour costs as thinly as possible. The result is a labour regime predicated on flexibility, driving employment practices which create precarity for workers. Their vulnerability is heightened by marginalisation which comes with being a migrant worker and intensified further for women and workers of colour. The result is a global horticultural workforce which is severely undervalued; a minimal portion of the financial value of produce is returned to them, for jobs conferred lower social status, in conditions predicated on precarity. UK horticulture is shaped by and contributes to this global context, as detailed in the next chapter. In this context, challenges faced by individual growers, such as the difficulty of offering good permanent jobs, arise from dynamics well beyond their control. Problems often regarded as the fault of individuals are far from being such.

There are particular knowledge dimensions to the current food regime, as corportisation shapes the skills ecosystem, and growers with narrow margins struggle to invest in knowledge exchange. Corporate control of agricultural inputs and associated R&D concentrates knowledge with companies that are then able to charge for the expertise required to grow crops. How this affects farmers' power over production has been well explored, with food sovereignty movements highlighting the fundamental right to access food growing knowledge. Knowledge at the hands-on level of food growing is less often emphasised, but equally significant as social constructions of "unskilled work" feed into the marginalisation of horticultural workers, making ideas about skill part of global horticulture's labour regime. Work portrayed as un- or low-skilled is subject to lower pay and protection, on the assumption that if it takes no skill, anyone can do it, making employees highly replaceable. As explored in the previous chapter, this pattern is apparent across industries associated with manual work, with the false categorisation of roles as unskilled serving to keep technical or practical jobs and associated educational pathways on lower rungs of employment hierarchies. Examples referenced here from across the world and the past century show how horticultural work has been positioned as lower status, with employers labelling their workers as unskilled to keep wages down or to justify exclusions from labour regulation. The concentration of women, migrants and people of colour in "low skilled" food production work has meant institutions have been more willing to downplay the rewards and protections due to workers as they have not been regarded as the core workforce, or main household earners, or were denied the humanity afforded to those racialised as white. These dynamics also play in the other direction: roles dominated by workers other than white male citizens are more likely to be regarded as lower skilled, and jobs which are low paid are assumed to be unskilled.

The marginal status of those most likely to do horticultural work is therefore deeply entangled with the imagination of their work as unskilled, a dynamic intensified by the food-migration nexus. The assumption that seasonal workers are low skilled excludes them from migration quotas favouring high-skilled immigrants, so they are forced to leave at season's end. These narratives feed misunderstandings about the reality of horticultural work, in turn contributing to a lack of interest in careers and training in the sector. Marginalised migrant workers are vulnerable to exploitation and abuse, and correcting this injustice is, of course, the immediate priority. There are also wider repercussions: exploitation tarnishes the image of horticultural work, making it even harder to attract recruits or address skills challenges.

As a focus on agri-food regimes indicates, state institutions are complicit in devaluing horticultural work. The cheap food regime is doing exactly what it is meant to and is highly adept at hiding the costs which fall disproportionately on producers and their workers. But when problems come to light, such as cases of on-farm abuse, systemic causes are rarely blamed. Flaws are located with individuals to mask the broader failings and injustices from which they arose, reinforcing the tendency to individualise how we imagine the work and skills of food growing.

3 The roots of "unskilled" work in UK horticulture

Introduction

The global context outlined in the previous chapter shaped UK horticulture, so its workforce reflects dynamics of the global agri-food regime; this chapter characterises the industry, tracing its roots through recent history. It reveals deeply rooted reliance on migration, and that horticultural roles have long been associated with groups marginalised in the labour market. Horticultural work and workers are not highly prominent in the history of UK agriculture or farm work. I therefore draw on material on agricultural work more broadly to provide historic context. Where sources and discussion relate to this, I refer to agriculture, farms and agricultural labour, specifying horticultural work only where distinguished in the source. To trace the roots of the industry's current form, let's first zoom into one region.

Roots in a region

The Vale of Evesham has a historic connection with horticulture, possibly since the days of its Benedictine Abbey, and remains a market gardening hub. I focus on the region here not only because earlier rural studies provide useful data, but for more personal reasons: I was born in Evesham and grew up in a village on the edge of the Vale. Our garden was surrounded by plum orchards; my school bus passed sprout fields and glasshouses, small fields with little more than a shack in one corner. As I read about the evolution of the region's horticulture, it connected with childhood memories.

No one in my family had a connection with horticulture and, beyond a token wigwam of runner beans, my parents weren't home gardeners. But fruit and veg growing seeped into our lives as they went from city incomers to village people. Our kitchen looked onto a quiet minor road, which around Easter became busier with cars and buses. Dad would adopt his "best" Brummie accent and imitate visitors marvelling at the views: "Innit lovely Doris? Ooh look at those hedges". These drivers were following AA tourist signs for the Blossom Trail, routed around the Vale to admire orchards in flower – a tradition apparently dating to just after the war (Buchanan 1948).

DOI: 10.4324/9781003281689-3

Orchards along the Trail were planted after the railway came to Evesham in 1851, a boom time when many local farmers switched from cereals to horticulture (Seabourne 2023). Before then apples and pears were grown for personal consumption or to pay labourers in cider and perry. One pear orchard was owned by a village family who needed extra hands for harvest. Mum collected me from the village school in her break from picking, taking me back to the orchard to work a few more hours, a large wicker basket looped onto a thick leather belt as she went up a wooden step ladder to the fruit. This time of year a couple of new children might appear in class, soon gone again without comment. They were from gypsy traveller families following harvest work, along peripatetic paths which used to be a mainstay of horticulture.

By my childhood in the early 1980s few orchards around the village were significant commercial operations. Some neighbours filled trays of damsons or plums to take to market, but it was not their main income; others picked for home use or didn't bother. I never saw the cider press which gave its name to a track down the hill in use. Parents of one school friend ran a plant nursery; no one else I knew worked in horticulture. Driving into town we passed minibuses parked on the edge of brassica fields, waiting to take groups home. I picture men in shalwar kameez and women in saris, I think a memory planted by later learning that gangs of workers were brought from Birmingham.

One summer holiday my sister cycled off with a friend to work as fruit pickers, joining decades of casual workers across the Vale. She never did a second day; her complaints deterred me from any first day. Some summers later, driving back to visit my parents I passed a couple in their early twenties, blonde and tanned, thumbs out, tramping along the verge. They jumped in the car and chatted about pea picking during summer break from university in Poland, exploring on a day off. I doubt there are many like them this summer – peas need picking but the pickers are now more professional.

Market gardens which bordered my school playing fields have been built on or converted to horseyculture, as smaller growers folded or were swallowed up. Bigger growers are easily missed beyond discrete signs with bucolic names such as Church Farm, tall trees screening static caravans from the road. But huge, refrigerated lorries on the small roads hint at the operation at the end of the drive. On summer evenings maybe you'll spot a bus load of workers being taken to the supermarket, bringing back beers to enjoy at their accommodation, instead of braving a village pub.

Horticulture in the Vale today is fairly typical of the UK's industry – significantly depleted and highly stratified, with remaining growers surviving through consolidation and specialisation. There are still glasshouses, growing tomatoes and salads for the mass market, operating as vertically integrated businesses focused on high-value crops (EFFP 2011). Many proudly describe themselves as family businesses, having evolved out of smaller farms. Other growers are oriented to alternative outlets such as farm shops, one of which I loyally swear sells the world's best asparagus. Annual celebrations like the Pershore Plum Festival and Asparagus Week signal a food movement absent during my youth.

Traditional orchards are included in the county's Biodiversity Action Plan, bringing funds for restoration. Where my mum picked pears, some trees remain, interspersed with camping plots, the fruit now more decoration than product. This spring when I turned into the village, the yellow sign for the Blossom Trail was in its annual spot, perhaps in hope more than promise, as visitors are more likely to see a new house called Orchard View than an actual view of an orchard.

The origins of the area's market gardening tradition are unknown (Sidwell 1969), but likely associated with its fertile soil (Ilbery 1985b). But human factors also fuelled growth, both through the contribution of specific experts (Sidwell 1969) and a unique land tenure system. The Evesham Custom established in the 1880s retains tenants' rights over improvements they make to land, incentivising investment in and care for a site, whilst enabling tenants to reclaim capital at tenancy end (Buchanan 1948). This encouraged farmers to diversify into horticulture when cereal prices were poor (Robinson 1983). The local council also made good use of the 1908 Smallholdings & Allotment Act to purchase holdings they made into small market gardens.

Production peaked in the 1940s, having grown alongside urbanisation and rail connectivity. The boom at the turn of the 20th century brought people into the area (Sparrow 2011). In the inter-war period, the Vale was the country's prime asparagus producer, with bundles going to Covent Garden. But decline began post-WWII, with approximately two thirds of farms gone by the 1980s. Most of those lost were tenanted small holdings as the council sold land, and others amalgamated (Ilbery 1985b). For many, low profits made it too difficult to invest in modernisation. Alternatives to hand cultivation were costly and unfeasible on small holdings. In the second half of the 20th century, land under fruit and veg production reduced significantly as transport costs fell, and the area lost its advantage to producers in Eastern England suited to modernisation (Ilbery 1985a). By the 1980s growers were small compared to the national average, but intensive, specialised and reliant on family labour. European competition favoured growers further east where industrialised production and processing were concentrated (Ilbery and Bowler 1994; Ilbery 1985b). This competition drove consolidation and specialisation in the Vale which continued into the 21st century (EFFP 2011). By the 1970s it had become much harder to make a living as a market gardener, so many sites became derelict as long hours and precarious incomes deterred potential successors (Sparrow 2011).

Ilbery 1985a Ilbery concluded:

> A vicious circle develops where reduced profits lead to uncertainty, a lack of capital for investment and labour problems; these in turn cause even lower profits and structural problems. The circle continues as a reduction in output creates yet less profit and capital for structural improvement.
>
> (1985b, 117)

Almost 40 years later, growers report similar challenges. Despite this, more growers remained in the 1980s in the Vale than would be expected to survive, perhaps due to attachment to a traditional way of life (Ilbery 1985a, 129). Ilbery found that those who carried on growing valued their job for a sense of independence and worthwhile work (Ilbery 1985a). These sentiments also live on, with many growers describing their pleasurable way of life, despite the hard work; the rural idyll remains convincing.

By the 18th century, most of Evesham's population were employed in gardening, meaning high winter unemployment portrayed as a financial and social problem, with the parish seeking to feed and contain the behaviour of labourers regarded as intellectually lacking (Sidwell 1969). But by the 20th century the Vale had a reputation for skilled market gardeners, using little more than a two-tined fork to conjure green gold from heavy soil (Buchanan 1948; Sparrow 2011). Growers were described as having "hereditary skill" and "intense pride" in their gardening, plus "a spirit of enterprise far above that of the average farmer and a rugged individualism which gives to the small grower immense staying power under adverse conditions" (Buchanan 1948, 250). As important were their wives and children who helped in fields and packhouses, their labour essential for the area to become a pre-eminent asparagus producer (Robinson 1983). Post-war, holdings remained labour intensive, employing between two and 30 workers. European migrants played a central role in production, with Italian entrepreneurs establishing market gardens after WWII (Sparrow 2011). The region's horticulture now relies on seasonal and non-UK resident employees (EFFP 2011).

So the Vale of Evesham has a special history of horticulture, which evolved in ways reflecting broader national trends. It is bifurcated into small-scale growers reliant on family labour and larger businesses employing many casual workers. Secondly, the price-cost squeeze drives growers to specialise, consolidate and industrialise, or exit. Third, the good life of being a grower is idealised whilst held alongside popular derision or dismissal of garden workers, and a reality of hard work often done by migrant workers. Next we zoom out to examine these patterns across UK horticulture, and how this shapes the sector's workforce.

UK horticulture: Grow big or grow home

It is almost 30 years since the death knell was sounded for UK horticulture facing gloomy prospects in the face of EU competition (Ilbery and Bowler 1995). But fruit and vegetable production remains a significant industry, albeit a minority within agriculture: it occupies less than 1% of agricultural land and generates almost a quarter of national crop value (DEFRA 2022a). Although there has been a steady increase in horticultural land over the last 15 years, coverage is 79% of 1985 levels (DEFRA 2022d). On land under edible horticulture, field vegetables dominate (75%). The value of domestic fruit and vegetable production in 2021 was £2,585 million (DEFRA 2022b).

There are fewer than 4000 horticultural businesses in England, of which almost 27% occupy less than 5 ha (DEFRA 2022c). In 2021, just over 1,000 horticultural businesses paid the statutory levy, meaning sales exceeded

£60,000/yr. Production in the UK contributed 57% of vegetables consumed nationally, whilst domestic fruit production contributed 15% (DEFRA 2022b). Less than 5% of vegetables grown in the UK (by weight, excluding potatoes) were exported, the majority to EU countries. In regions which have retained concentrations of horticultural producers, the sector is seen to offer good economic potential associated with demand for local produce (Worcestershire LEP 2016). But growers have been affected by recent pressures: the impacts of Brexit, the Covid-19 pandemic, inflation, particularly in energy prices. During 2022 a steady stream of growers reported winding up or leaving spaces unplanted to avoid uneconomic production; a House of Lords inquiry found prospects for the sector incredibly bleak (Horticultural Sector Committee 2023).

UK horticulture is characterised by the bifurcation apparent in the Vale of Evesham since the mid-20th century. Around 90% of edible horticulture producers are classified as micro-businesses (Pye Tait 2020), a government defined category for those employing fewer than ten staff or with a turnover of less than £632,000. At the other extreme are large multi-site operations, some with vertically integrated businesses incorporating processing or packaging, which are amongst UK agriculture's biggest employers. This creates hugely divergent employment profiles. Modelling based on a recent survey calculated the most common size of staff teams in edible horticulture as two, with 25 businesses employing more than half the sector's staff, having an average workforce of 1500 (Pye Tait 2020). As detailed in the previous chapter, globalised agri-food supply chains pressure producers to consolidate, specialise and scale up, creating large highly specialised businesses. Production around the Mediterranean enjoys a more favourable climate and lower labour costs, against which UK producers struggle to compete, creating a cycle of low profits, and limited capital to invest in enhancing profitability (Ilbery and Bowler 1994).

The resulting bifurcation is particularly apparent in the UK's recent boom crop – strawberries. From the mid-1990s multispan, open sided Spanish polytunnels became popular, bringing longer growing seasons and a more reliable crop (Calleja et al. 2012). Domestic production became more competitive, leading to specialisation and considerable investment in infrastructure which smaller or non-specialist farms could not afford. The rise of supermarkets brought price pressure which undermined pick-your-own and wholesale markets supplied by smaller growers, resulting in their demise. Similar trajectories occurred across UK horticulture:

> Productivist practices have become increasingly popular and are almost seen as being inevitable by the industry, leading to the intensification, specialisation and concentration of horticultural practices. These are believed to have been driven by supermarket dominance of the fresh food sector, as they drive prices down and force farmers to maximise their output at the lowest relative cost.
>
> (Calleja et al. 2012, 609)

The low-price imperative forces the decision to "grow big or go home" – scale up or go out of business, alternatively focus on markets close to home serving quality-driven consumers.

In the previous chapter, we saw how globalised supply chains disempower producers relative to retailers. The economics of this for UK producers are apparent from analysing the price consumers pay (Jack and Hammans 2022). For UK grown apples, 1 kg of dessert apple retailing at £2.20 in a UK supermarket generated 8p profit, which is distributed equally between the retailer, the wholesaler and the grower, but the grower bears 35% of costs, meaning they earn only 3p per kilo. Crops sold as loss-leaders in supermarkets have harsher economics for growers. For example, carrots retailing at 54p/kg generate negligible profit for growers who receive around 14p/kg, roughly equalling break-even; and growers report pressure to sell at less than cost price (Horticultural Sector Committee 2023). That said, some emphasised to me their positive, long-term relationships with supermarkets. It is also worth noting that alternative retail is not without issue, as local, independent supply chains also face pressure to keep prices low. Smaller, agro-ecological growers often face the most acute economic pressures (Dumont and Baret 2017). Pressure to produce cheaply is endemic in globalised food systems as this stakeholder described:

> The supermarkets are part of the problem for the fresh produce industry, but they do the job that the government wants. The government wants cheap food so that's what the supermarkets do. One of the whole problems is food is too cheap, so producers are being driven into the ground all the time
>
> (Sh16)

Forces pressing horticultural producers towards low-cost work regimes run deep and wide.

It's a dirty job but someone's got to do it – The UK's horticultural workforce

It is difficult to present an accurate picture of who works in UK horticulture, or the extent of labour shortages. The government reported a total of 34,000 people employed in horticulture in England during 2021, around 11% of the total agricultural workforce (DEFRA 2022c and 2022d); others calculate a much larger workforce. A survey of businesses reflecting the industry profile provides one of the most detailed pictures of horticulture's workforce, although perhaps influenced by a period of uncertainty regarding new immigration rules and early impacts of the Covid-19 pandemic (Pye Tait 2020). This characterised edible horticulture's workforce:

- 100–120,000 employees.
- Businesses employ between 2 and 250+ with the most common number of employees being 2.

- The largest 30% of businesses employ around 90% of the workforce.
- 60% are permanent staff and 40% seasonal.
- 72% are in non-specialist roles, defined as not having a technical occupation, but may have skills or certified competences (e.g. forklift driving, spraying).
- 51% of employees are UK citizens and 47% EU citizens, of which 42% are Bulgarian, 28% Romanian and 17% Polish.
- Vacancies for managerial and supervisor roles represent 19% of the workforce at that level, a little higher than the national average for any industry.
- 11,569 seasonal vacancies were reported.

Levels of vacant roles reflect a broader problem with recruitment across agrifood industries (Independent Review into Labour Shortages in the Food Supply Chain 2023). Horticultural actors have been suggesting for more than a decade that they face a skills crisis, with shortages across all levels now suggested to be chronic (Horticultural Sector Committee 2023). But the precise shape and scale of these shortages is difficult to confirm, with skills and labour issues often conflated, and a lack of robust data (Pitt 2021b). The sector's reliance on large numbers of temporary workers makes it particularly difficult to track the labour force, with no official data monitoring (Office for National Statistics 2019). In 2021, 38% of horticultural workers were employed on a casual basis, the highest proportion across agriculture by a significant margin (DEFRA 2022c).

Work across UK horticulture is dominated by people born overseas: more than 90% of UK horticulture's seasonal labour are migrants from the EU (McGuinness and Garton Grimwood 2017). Significant numbers initially came from other EU states as horticultural workers, then settled as permanent residents, indicating the sector's reliance on the food-migration nexus. As our zoom into the Vale of Evesham revealed, this is not a recent phenomenon as post-war settlers provided agricultural labour and established market gardens (Sparrow 2011). The UK's first scheme bringing seasonal workers into horticulture began in 1945 (Davies 2022); the industry's reliance on migrant workers has since increased significantly, in parallel with broader European trends (Corrado et al. 2017b; Gertel and Sippel 2014; Rye and Scott 2018).

The government's last survey of seasonal labour showed that in 2018:

- 35% of horticulture farms used seasonal labour,
- larger farms were more likely to use seasonal labour – 55% compared with 18% of small farms,
- 48% of those using seasonal labour reported that at least three quarters of workers live in the UK year-round,
- 33% of horticulture farms that used seasonal labour experienced a shortfall of seasonal workers,
- seasonal workers' average hourly pay was £9.15 including any piece rates (minimum wage at the time was £7.83/hr) and
- 28% of horticulture farms that used seasonal labour paid less than £8.00/hr, whilst 12% paid over £11.00/hr (Defra 2019).

In the last two decades of the UK's membership of the EU, most seasonal workers came from recently acceded states; the majority now come from Bulgaria, Romania and Poland (Pye Tait 2020). Since the reopening of a seasonal workers scheme in 2018, the proportion recruited through this has increased to around 60% (NFU 2022).

Although some seasonal workers occupy managerial or supervisory roles, the majority could be classified as non-specialised or non-technical, mostly harvest roles (Pye Tait 2020). Soft fruit growers need large teams of pickers for almost nine months of the year, whilst tree fruit such as apples and pears have a shorter season, meaning just a few weeks of harvest work. Harvest jobs do not suit those needing a year-round income, and as rural housing has become more expensive, many living near large growers could not afford to rely on them (Nye and Lobley 2021). Many growers provide on-site accommodation, such as caravans, to avoid expensive rural rents or commuting; cheap housing available in the south coast's faded seaside towns has added to the region's advantages for intensive horticulture. The temporality and location of jobs in horticulture require workers with mobility and itinerancy: short-term roles, living elsewhere off-season. Before exploring the resulting seasonal workforce, it is important to consider its historic context as a long view on horticultural work highlights that today's trends are far from new; agricultural work has long been regarded as undesirable and unskilled.

Roots in the past – A history of devalued work

> If low pay has been a constant in the history of the farmworker, their place in the social and economic landscape of rural England has become more and more peripheral.
>
> (Verdon 2017, 244)

The UK was amongst the first countries to shrink its agricultural workforce, as across the 19th century cities and industrialisation spread, and farming mechanised and consolidated. The Industrial Revolution and enclosure movement focused UK agriculture on maximising production for the market, as population growth fuelled demand and technical innovations powered productivity (Martin 2000). As noted in the previous chapter, during this first global food regime, the UK relied on cheap imports, particularly from the colonies (Zerbe 2019). Agricultural work and workers became more marginalised, largely absent from public consciousness, until moments of crisis foregrounded them (Verdon 2017). Farmworkers remained low paid and were typically perceived as unintelligent, under-educated and socially backward (Verdon 2017, 2).

In the late 19th century, agricultural profits and confidence were high, with horticultural crops benefiting from price rises driven by urban demand (Verdon 2017). Whilst the rest of agriculture reduced its workforce and offered permanence with slightly better incomes, expanding market gardens needed workers, particularly for seasonal tasks like harvesting. Regions growing more plant crops relied on casual workers from the surrounding area, but also from some distance, particularly for harvest season. For example, Irish men would leave their livestock farms for hay making. Lincolnshire's horticultural farms grew to supply consumers in large urban areas like London and Sheffield, with harvest workers moving in the opposite direction (Verdon 2017). Some harvest work like pea and fruit picking around Maldon, Essex suited women and children from the region. More labour-intensive harvests like hops drew workers from further afield, providing seasonal work for poorer urban residents. Herefordshire drew workers from the Black Country; their Kent counterparts drew workers from London to work alongside local men and women, a demand which continued into the new century, with more than 100,000 pickers required for the 1908 harvest (Verdon 2017, 60). These influxes of city folk required special lodgings and were barely tolerated by the rural communities who saw them as disruptive outsiders. Data does not accurately report numbers involved in casual farm work, but testimonials make clear the significant role of women and children, even post-1880 when compulsory education officially took them out of the fields (Verdon 2017).

Arable and horticultural production which by the late 19th century was dominating East Anglia meant the region required a significant agricultural labour force for specific periods. The solution was a gang system for hiring and organising labour, introducing intermediaries between employer and worker, with a sub-contractor engaged to provide sufficient labour for a task. The system suited cheap food production, and declining provision of poor relief (Strauss 2013). The gangmaster was typically paid for completion of the task, creating incentives to reduce workers' pay and conditions (Verdon 2017, 97). Gangs often included women and children, working under tough conditions and walking some way to work. These practices prompted a series of commissions investigating agricultural labour, which voiced concern about work conditions and gang's moral impacts on "female virtues" (Strauss 2013). A licence for gangmasters was established in 1867, which Verdon argues was driven more by these moral concerns than any will to tackle the system's inherent economic exploitation (2017, 100).

The gang system is not the only feature of farm labour relations established during the 19th century which echoes through to today. Firstly, employers and farm owners had the upper hand in hiring practices and employment conditions as there was no shortage of willing workers. The exception was during harvest when the importance of timely work and uplift in demand reduced farmers' room to negotiate (Verdon 2017, 72). Even as the government began to intervene in other aspects of agriculture, they remained reluctant to intervene in the labour market and introduced only weak controls on practices such as gangmastering. Secondly, pay and social standing associated with farm work varied according to the role, with permanent, full-time male workers best paid. Specialisation was rewarded, with men working with animals perceived as the most skilled and valuable workers (Verdon 2017, 65). Women and seasonal workers were paid less, and the tasks they dominated were seen as less skilled (Verdon 2019). Thirdly, production of fruit and vegetables relied on workers available for short intensive periods, demand met through casual employment of local women and children, seasonal migration of the urban poor and longer migration by Irish and gypsy traveller men. Seasonal influxes to rural communities were tolerated as necessary, but resented, creating insider-outsider dynamics (Verdon 2017). Finally, work done on farms was perceived, particularly by powerful urban classes, as inferior to other work in terms of skill:

> Farm work was depicted as a formulaic, unvarying and mind-numbing occupation. Its workers were low paid and low skilled, and were destined to remain so; they lacked prospects for advancement, devoid of the means or ability to progress.
>
> (Verdon 2019, 831)

As Verdon explores, there were moves to counter this perception with, for example, government commissions considering the learning and application of skills in farm work, but this did not extend to casual workers or roles typically done by women (2019). Given that both these groups were prominent in horticultural roles, it is clear this work was more likely to be deemed unskilled. In the Victorian era, tasks associated with "concentration and dexterity rather than skill or training" were seen as suitable for women and children rather than more professional male farmworkers (Verdon 2017, 52). Weeding, picking stones or sorting crops were said to favour women's nimble fingers, whilst children's short backs were appreciated for tasks low to the ground (ibid.). Verdon reveals assumptions about age and gender which shaped segregation of task allocation and reward structures; the women working in Victorian England's fields were regarded in similar ways to those in Californian salad packing, or today's strawberry packhouses. Their supposed feminine traits made them good workers, but not professional ones, as they really belonged in domestic spaces where their more important role was to feed men doing skilled farm work.

Crises of the 20th century

The inter-war period was a good one for horticulture with intensive producers across Middlesex, Bedfordshire, Essex, Hertfordshire, Gloucestershire, Herefordshire, Kent, Shropshire, Surrey, Sussex and Worcestershire having healthy outputs supporting year-round jobs (Verdon 2017, 176). Intensive potato production in Lincolnshire and Cambridgeshire also employed a core of regular workers. Seasonal fruit work in Gloucestershire and Kent and nursery work in Hertfordshire used "long-term casuals", women from surrounding villages returning regularly to the same farms, whilst East Anglia continued to rely on the gang system. The annual exodus from London to

Kent's hop harvest continued, with the Black Country providing summer workers to its rural periphery and Nottinghamshire colliery towns supplying that region's pea pickers.

The first half of the century brought a global agricultural crisis due to the World Wars and Depression, with serious repercussions for farm work. These have interesting parallels with more recent shocks to horticultural work, with four notable trends reverberating through to the 21st century. First came public appreciation of the importance of domestic agriculture for food security, so the government executed a dramatic policy U-turn to shore up production (Martin 2000). The prospect of farmworker shortages drove new public intervention to oversee labour supply (Verdon 2017, 136). This period brought significant public investment in the agricultural knowledge base through support for training in agricultural work and R&D funding, as indicated by the creation of the Agricultural Research Council (ARC) in 1931. The ARC was intended to coordinate agricultural science and successfully increased productivity (Cooke 1981). However, new knowledge did not always reach farmers, with low uptake of publicly funded places on agricultural courses (Martin 2000, 23). Martin suggests this was due to suspicion of formal education within agricultural communities who mostly learnt on the job (2000, 23). This is another trend apparent today: negative feedback loops where lack of demand for courses reduces investment in the infrastructure for agricultural knowledge exchange, and a workforce reliant on informal learning.

The third trend with parallels today was how fear of labour shortages altered workers' status. Shortages prompted steps to enhance agricultural labour conditions, both through government intervention and farmworkers' self-organisation as they became more aware of their significance and rights (Verdon 2017). The Agricultural Wages Board for England and Wales was created in 1922, with powers to enforce minimum wages and work conditions. The Board included representation for farm unions who were gaining members and influence. However, average wages remained low relative to other industries, and farmworkers were excluded from new protections such as National Insurance and statutory holiday requirements (Verdon 2017).

During both wars, the threat of vacancies left by men going to fight prompted employment through non-typical routes. This included using soldiers, prisoners, children and, most famously, women of the Land Army – although their contribution has been exaggerated and was not universally welcomed at the time (Ginn 2012; Verdon 2017). Harvest work was done by volunteers from urban areas who stayed in camps designed to foster rural-urban cohesion (Verdon 2017, 215). Understanding the exact extent of wartime worker shortages and the significance of novel recruitment is complicated by the likelihood that it was in the interest of actors like AWB to exaggerate shortages (Verdon 2017). This construction of labour shortages lives on in today's narratives (Chapter 4).

A fourth historic trend resonates with the discourse around the Covid-19 pandemic: the celebration of food workers during crisis. A strong public narrative around the need to feed the nation enhanced public perceptions of agricultural work, giving farmworkers key wartime roles "both practically and symbolically" (Verdon 2017, 191). However, any shift in perceptions was driven by fear of hunger, rather than closer appreciation of the work itself. Celebration of agricultural work focused on novel workers such as the Land Army more than those who were the mainstay of agricultural production and did not persist (Verdon 2017, 192). When economic conditions deteriorated after WWI, the government disbanded the AWB. The average agricultural worker still lacked prospects of upward mobility through owning their own farm, or moving into better paid industries (Martin 2000, 19). Lower than average wages persisted and deterred people from entering the industry (Verdon 2017, 225-6). Other workers regarded farm labourers as low-skilled, giving them low levels of esteem which deterred young recruits (Armstrong 1988, 201). These patterns suggest that how British people and policy makers act towards agricultural work and workers during a crisis is exceptional, rather than prompting enduring attitude change.

Productivism and "deskilling"

As outlined in the previous chapter, the first half of the 20th century marked the transition between the first and second global food regimes. Agriculture rapidly modernised thanks to state investment in intensification, representing the UK's second agricultural revolution as productivity increased dramatically through industrialisation (Healy et al. 1985). Production reliant on expensive technology and inputs such as fertiliser requires capital, driving consolidation and specialisation. Farm numbers declined whilst farm size increased creating large, capital-intensive, specialised farms (Healey et al. 1985). Increased prominence of machinery brought some respect to farm work, but also negative perceptions of agricultural work as many jobs became isolated (Martin 2000, 107). Farmworkers' skills remained undervalued because pay structures tied skill level to remuneration, making employers disinclined to recognise employees as skilled (Verdon 2017, 239).

Horticulture's protected crop production in glasshouses faced considerable pressure to scale up because of its capital intensity (Phillips 1985). The 1970s oil crises made inputs dependent on fossil fuels – energy, machinery, chemicals and transport – more expensive, whilst prices for resource-intensive crops like tomatoes stagnated or fell. Maximising productivity was paramount, and government policy at the time favoured self-sufficiency, meaning significant grants for better, larger glasshouses (Phillips 1985). Those who took this opportunity formed larger businesses or worked together, creating scale and efficiency which could survive cost-price squeezes and deal with the increasingly dominant supermarkets. By 1985 the small-scale market gardener was "producing at high cost for a shrinking market" (Phillips 1985, 123). The corporate food regime shifted power from producers to retailers, driving consolidation and specialisation, with only the largest, most efficient growers able to supply supermarkets, particularly in the face of imports from rapidly expanding Mediterranean production.

Technological innovation was enabled by public investment in specialist research institutes (Martin 2000, 81). This is often regarded as the heyday of the UK's horticultural knowledge system, with specialist centres such as the National Vegetable Research Station established in 1949. Expertise in plant breeding was nationally funded through institutions like the National Institute of Agricultural Botany (Martin 2000, 99). During the 1970s, UK government was spending £4.1 million annually on research in glasshouse crops alone (Phillips 1985, 125). R&D capacity developed alongside expansion of training and technical advice services promoting farm innovation (Martin 2000, 91).

These transformations required less human labour, so numbers of farmworkers declined significantly through to the 1980s. Agricultural work continued to change as farms became more specialised and workers focused on a smaller range of more complex tasks (Healey et al. 1985). The competitive economic environment drove many farm owners to seek incomes beyond agriculture, driving the rise of the part-time farmer (ibid.). Increased use of technology, innovation and specialisation meant agricultural work became more managerial than manual, with entrepreneurial skills becoming important (Bowler 1985). Changes were often interpreted as a form of deskilling as roles became more repetitive, and knowledge was outsourced to companies producing inputs (Morgan and Murdoch 2000; Timmermann and Felix 2015). As individual farmers apparently needed fewer skills, they seemed more replaceable and lost bargaining power (Timmermann and Felix 2015).

Although it is questionable that farmers truly lost skills, or stopped developing specialist knowledge (Oreszczyn et al. 2010), companies which develop inputs became a considerable feature of agricultural knowledge. As they tend to commodify and control the knowledge they develop (Morgan and Murdoch 2000), this established a risky context for the phase of public disinvestment which was to follow. By the 1970s horticultural research in the UK had reached peak investment but was soon to suffer reduced government support (Pollock, 2012). Specialist institutes were privatised and consolidated, as the government sought to separate applied and strategic research, with the industry expected to bear the cost of the former. The loss or privatisation of institutions focused on R&D closer to production left a weaker, less strategic knowledge pipeline, with specialist R&D more fragmented and dependent on commercial investment (Brian Jamieson and Assoc. 2008). This is particularly problematic for horticulture where there are so many crops to be investigated and relatively few producers specialising in each (Menary et al. 2019). A largely privatised system also suffers trust issues as producers are sceptical of advice from market-driven providers (Curry et al. 2012). These concerns led producers to vote for a compulsory levy to fund R&D, resulting in creation of the Horticulture Development Company (Gill 1988), which evolved into the Agriculture and Horticulture Development Board (AHDB). By the tail end of the 20th century, specialist horticultural research was largely a private sector concern, with reduced public investment and oversight leaving a depleted, fragmented infrastructure reliant on commercial interests.

Horticulture's labour intensity was also in decline, as apparent from the example of tomato growing. In 1976, 57.6 "man hours" (sic) went into producing a tonne of tomatoes, which had fallen to 41.1 by 1982 (Phillips 1985, 127). However, harvesting remained labour intensive meaning continued reliance on casual workers, often women. Migration routes opened during the war, such as summer camps, continued and until 1951 were supplemented by displaced people from Central and Eastern Europe (Verdon 2017). The shrinking pool of farmworkers undermined their status and pay, with horticultural roles particularly vulnerable as they included considerable numbers of non-technical roles fulfilled by migrants and women, placing them at the bottom of hierarchies of agricultural workers.

Into the EU

The UK's entry to the EU in 1973 did not mark a significant switch in policy as productivity and increasing farm incomes through market interventions retained priority (Martin 2000). Whilst the Common Agricultural Policy (CAP) had some impact on farm incomes, the benefits tended to favour those with higher incomes; labourers were still low paid (Martin 2000, 149). Horticulture received the least direct EU financial support (Ilbery and Bowler 1994) and saw marked consolidation as smaller growers went out of production. The common market put UK growers in competition with those on the continent with more favourable conditions. Southern European states had lower energy and wage costs, whilst those in Northern Europe received more government investment (Ilbery 1985). This drove a sustained period of decline in UK production, with intensive production concentrating in regions suiting to large, highly specialised, vertically integrated enterprises; in East Anglia and the Southeast, levels of sunshine and proximity to urban markets allowed horticulture to thrive (Ilbery and Bowler 1994). The feature of EU membership with the greatest impact on UK horticulture was the right to live and work in any member state.

Over here, over skilled and underpaid

As noted in Chapter 1, horticulture globally is reliant on migrant workers, which heightens risks of exploitation and devalued work. Seen in their historic context, perceptions which undervalue these workers' abilities and disregard their rights are not an aberration, rather the culmination of deep-rooted patterns. Those growing the UK's produce have always included migrants: growers still active in the Vale of Evesham include descendants of Italian prisoners brought to work during World War II, and families who migrated from the continent to escape post-war turmoil (Sparrow 2011). Unlike other 20th century agriculture, horticulture required a large labour force, mostly for manual, repetitive, demanding roles in intensive environments (Verdon 2017, 253). Local women who had historically been a mainstay of seasonal roles found these jobs less attractive, so the industry looked to the spatial fix of seasonal migrant labour (Scott 2013). Migrants' willingness to locate temporarily in rural areas suited horticulture's appetite for large numbers of seasonal workers. The first wave came from relatively poor Southern European countries like Portugal, followed in the 1990s by those from the Eastern bloc, then from countries acceding in 2004 (Verdon 2017, 254), as economic disparity between member states shifted.

The Seasonal Agricultural Workers Scheme (SAWS) established in 1990 set a quota of migrants entering to work in agriculture, with most taking jobs in horticulture at harvest time. SAWS was conceived as a summer cultural experience for young students, but workers' profiles evolved, as did SAWS' scale and significance (McGuinnes and Garton Grimwood 2017), peaking in 2004 at 25,000 places (Scott 2022). By 2011 in a horticulture hotspot like the Midlands, for every UK citizen working in a seasonal role, there were seven from overseas (EFFP 2011). Despite significant industry lobbying for its retention, the government closed SAWS in 2014 in the expectation that the free movement of labour would make it obsolete.

Intra-EU migration did indeed bring workers into UK agriculture as poorer countries joined the common market and their citizens sought to earn in higher income member states. Many of the 1 million plus people who came to work in the UK between 2004 and 2011 from A8 states (Poland, Czech Republic, Latvia, Lithuania, Slovakia, Slovenia, Hungary and Estonia) worked in agriculture (Findlay & McCollum 2013). Later accession states – Bulgaria and Romania – went from relying on SAWS visas to gaining the right to work in the UK and soon dominated horticultural work. These demographics are apparent in today's workforce, as many who came temporarily settled and progressed through the business. I have visited many growers where managers or those in more technical roles include Poles or Lithuanians, the earlier migrants, whilst those who came later from A2 states are following them up the promotion ladder, having begun in seasonal posts latterly occupied by Ukrainians.

The government's justification for closing SAWS was caught up with perceptions of agricultural work as unskilled. It offered a rare route to work in the UK as an "unskilled" recruit, on the basis that horticultural work requires no specialist education or ability (McGuinness and Garton Grimwood 2017). As hostility to immigration grew, the government promoted a vision of a high-skill UK economy, prioritising educated and skilled immigrants, but horticulture and other agri-food businesses relied on migrants to fulfil "unskilled" roles. Agricultural employers were conscious that SAWS tied workers to them, whilst in the general migrant employment market, they would compete with more appealing entry-level occupations such as hospitality (Scott 2022). There are too few British workers willing to fill these roles (Nye and Lobley 2021). Moreover, seasonal workers are highly productive, with those who return for successive seasons valued by employers for their dedication and productivity.

One grower told me that when they had fewer experienced pickers during the pandemic, first-timers were only picking fast enough to produce £3/hr, meaning the remaining £5 to take them above minimum wage were costs absorbed by the company.

Anti-migration sentiments grew and played a significant part in the UK's 2016 vote to exit the EU. The vote initiated a period of incredible uncertainty for horticultural recruiters, as they urged seasonal workers from within the EU to return, despite lack of clarity regarding immigration rules. It also caused a drop in the pound's value making the UK a less attractive destination for migrants. By 2018–2019 uncertainty about the residential status of migrant workers in or planning to come to the UK, plus the prospect of increased controls on numbers and types of migrants, and anti-immigration rhetoric combined to create horticulture's Brexit crisis (Milbourne and Coulson 2021). Growers reported harder recruitment, but this crisis was at least partially constructed to allow the industry to argue for continued access to migrant workers (Milbourne and Coulson 2021). The committee tasked with indicating which sectors have vacancies, justifying consideration as "shortage occupations" deserving favourable migration rules, found no need to include farm or horticultural work (Migration Advisory Committee 2020). Eventually government conceded there was a labour problem and established a seasonal worker visa. Initially, numbers were very restricted (2,500) on a pilot basis, expanded for 2020 (10,000) but still inadequate to the industry's request for a larger, long-term scheme. Their pleas were met with an increase to 45,000 visas, a figure set to taper, as automation and domestic workers replace migrants.

The seasonal worker visa indicates UK horticulture's reliance on migrant workers, making it typical of the current global agri-food regime and the food-migration nexus. These workers bear the weight of the corporate need for flexible workers through multiple forms of precarity, sometimes abuse and exploitation. Seasonal workers in the UK are relatively well protected in comparison with counterparts elsewhere as all farmworkers are entitled to the national living wage; some receive additional piece rates, which earn fast pickers good incomes. European workers are better protected than those in the USA by labour regulation, and competition between countries drives up conditions (Hahamovich 2003). As seasonal worker availability declined across Europe, UK growers struggled to compete with countries offering more favourable exchange rates (Milbourne and Coulson 2021). This focused attention on retention, meaning it is in employers' interests to build a reputation for good pay and conditions (Pitt 2021a). But UK consumers cannot rest easy that they are not eating products of exploited labour. A majority of imported produce comes from EU countries, giving some guarantee of labour regulation, but there is systemic exploitation of horticultural workers across Europe (Corrado et al. 2017b; Sippel and Gertel 2014). The Netherlands from where the UK receives the largest proportion of its imports (by value) has well-regulated employment practices, but is far from fair in how it treats workers, with standard practices straying towards unfree labour (Siegmann et al. 2022). More distant exporters such as Morocco rely on intense employment for low pay (Medland 2021). In other African countries, food production includes exploitation invisible to consumers in the Global North (Greco 2020). The fetishistic veil capitalism throws over commodity supply chains is particularly effective in hiding remote exploitation.

Nor is home-grown produce free from exploitative labour practices, as pressure to produce as cheaply as possible pushes some in this direction (Schenner 2018; Strauss 2013). Many UK migrants are recruited via intermediary agencies or gangmasters, sometimes operating illegally, seeking financial gains from migrants who were not informed of their rights to minimum wage or protections from forced labour and debt bondage (Verdon 2017, 257). The period prior to gangmaster regulation intensified this, as revealed by Chinese cockle pickers drowning in Morecambe Bay in 2004 because they were not properly protected by the gangmaster who employed them illegally. The previous year, a minibus of horticultural workers travelling to work in the Vale of Evesham was hit by a train at a level crossing, killing three. Their employer was adamant all laws and welfare provisions had been met, but the Iraqi driver had no licence, and his inability to read safety instructions was declared a factor in the accident (Press Association 2004). Public reaction to these tragedies and other undocumented and trafficked workers led to the Gangmaster Licensing Act 2004, establishing a regulator and inspection regime intended to ensure licensed labour agents adhere to fair practices. However, enforcement is under-resourced and insufficiently stringent to prevent abuse of horticultural workers (FLEX 2021).

The independent organisation Focus on Labour Exploitation (FLEX) considers horticulture a sector with high risks of exploitation due to reliance on large numbers of migrants, isolated workplaces and supply chain economics. Its research with workers on the pilot seasonal worker visa identified significant risks of forced labour according to ILO criteria (FLEX 2021). The rapid expansion of the visa seems to have borne out their concerns as the investigation found abusive treatment on farms, and workers illegally charged to secure visas (Mellino and Pangeni 2023). Post-Brexit, fewer European workers are coming to the UK, with recent seasons also hit by the Russian invasion of Ukraine cutting off migration. Seasonal visas now go to people from former Soviet states and Southeast Asia recruited through new intermediary relationships which were perhaps established too rapidly. Recruits from Nepal and Indonesia were charged illegal fees to secure jobs, arriving in the UK to find no job, or not guaranteed sufficient work to earn back their debts (Dugan 2022). There have been concerns regarding compliance with scheme requirements, and lack of enforcement inspections. The Independent Chief Inspector of Borders and Immigration found inspectors had identified concerns with welfare and poor treatment of seasonal workers, but:

> Overall, the Home Office has not demonstrated that it has the mechanisms or capabilities in place to assure itself that scheme operators are meeting compliance requirements. When serious concerns have been raised by workers themselves, it did not act promptly or seriously.
>
> (ICIBI 2022: 1.25)

Rapid efforts to fill vacancies pushed agents to look further afield, relying on less established relationships with agents who are then blamed for wrongdoing. The longer the supply chain feeding recruitment, the more difficult it is to guarantee fair treatment and transparency. Supermarkets selling the fruit picked by exploited workers are separated from them by two or three intermediaries, although as more cases came to light, they became sufficiently concerned to establish new audits (Gonçalves 2023). But the scheme continues, albeit with a new requirement to guarantee hours of work. Growers exposed for mistreating workers are still advertising seasonal vacancies. The regulator is still under-resourced, and the conditions making horticultural workers vulnerable to abuse persist.

The rules and regulation of seasonal work visas have multiple problems, but do offer a legal route into horticultural jobs, and providing workers are protected and listened to, they can meet their needs (Scott 2013; Weiler and Mclaughlin 2019). But workers coming to the UK on seasonal visas are vulnerable because they are absent from the public realm. On-farm accommodation in isolated locations means the settled population rarely meets them (Lever and Milbourne 2017; Rye and Scott 2018). They never feature on packaging or adverts, whilst supermarkets like to celebrate the name and image of farm owners. In just food systems all workers should be valued (Sbicca 2018), but invisible workers are easily devalued (Brown and Getz 2011). And employers are disinclined to make them visible: "there is a tendency not to want to unveil the exploitative labour practices that have underpinned an era of relatively cheap and convenient food" (Scott and O'Reilly 2022, 3).

Employing migrant workers also hides social impacts and inequalities in their home country. One employer described to me how the pandemic was tough on their team as it was impossible to visit family at home. Several returnees usually left their children with their grandparents for the season, an arrangement which was impossible during the pandemic meaning some had to miss that year's work. All those summers away from their children to earn good money as the farm's fastest pickers, relying on reproductive labour back home, with who knows what impacts on the children and their grandparents. As the global horticultural workforce has feminised, such arrangements are common, with many women juggling the labour of social reproduction with waged work (Medland 2021); employers and government favour these women as they are more likely to return home (Hellio 2014; Nieto 2014).

Even when well-paid and treated, it is arguable that seasonal migrant work in UK horticulture cannot be just. The global extent of schemes like the seasonal visa shapes a labour regime bringing "new dimensions of insecurity and vulnerability" (Constance 2014, 163). In the previous chapter, I highlighted that growers, government and citizens like this arrangement because it provides workers without extending citizenship (Hahamovich 2003). Extracting labour without offering full rights of citizenship or permanent right to remain is a fundamentally inequitable exchange, with the migrant disadvantaged compared to domestic workers. Horticultural workers migrating within Europe and treated legally remain in precarious positions, underpinned by the regulatory system (Siegmann et al. 2022). Migrant flows are also inherently unjust because they rely on global inequalities which make it attractive for workers to move towards economic centres where they can earn higher wages. And these inequalities in turn depend on hugely unjust, exploitative colonial relations which extracted resources, rights and lives from colonised peoples and which persist in racial capitalism. Although these conditions could rightly be termed crisis, they are not the forms of crisis which horticulture in the UK has highlighted.

Crisis? What crisis?

As Verdon's history of farm work shows, the industry's workforce has occasionally benefited from moments of crisis which heighten attention to their importance. Actors have been able to parlay crisis narratives into special status for horticulture in immigration policy, as explored in detail in the next chapter. A feature of this narrative prominent since the turn of this century is significant here: horticulture's skills crisis. In 2013 the Royal Horticultural Society (RHS) convened a group focused on what they called horticulture's growing skills crisis. They found that 72% of surveyed businesses were unable to fill skilled vacancies; 70% of 18-year-olds believed horticultural careers were only for those who failed academically; nearly 50% of under-25s thought it an unskilled career (Horticulture Matters 2014). The campaign which followed was one of several initiatives seeking to address skills shortages (Pitt 2019 and 2021b). They seem to have made little headway, as a House of Lords committee concluded in 2023 that UK horticulture faces "a chronic shortage of workers at all skills levels" (2023, 43). The skills crisis is not always well defined, but centres on three connected problems. First is the inadequacy of education and training programmes for future growers. Whilst symptomatic of underinvestment in further education (FE) and the enduring divide between vocational and academic education, horticulture has fared particularly badly (Dixon 2016). Restructuring of further education focused on financially viable courses, resulting in fewer programmes focused on horticulture, especially production or edible horticulture (Aldous et al. 2014), as they are relatively expensive to deliver (Pitt 2019). It has also been suggested that horticultural businesses are particularly disengaged from training provisions, or even mistrust the programmes available (Strawbridge et al. 2011). A 2020 survey of growers found that almost a third are not accessing external training provision because they cannot access it easily or affordably (Pye Tait 2020). In higher education too, there are few courses focused on horticulture, weakening the pipeline of specialist producers and researchers (Pitt 2019).

Lack of educational courses is linked to the second feature of the skills crisis: reduced capacity in specialist R&D and associated knowledge exchange. Investment in horticultural R&D has declined since the 1990s, with many specialist institutions closing or becoming privatised (Brian Jamieson & Associates

2008). The remaining provision is fragmented and uncoordinated, with the market logic making actors defensive and uncollaborative, inhibiting innovation (Menary et al. 2019). Although not the only factor, it seems that dissatisfaction with research and its dissemination influenced horticultural producers who voted in 2021 to end the compulsory levy used to fund research and innovation.

To a degree, challenges around horticultural R&D reflect wider agricultural knowledge systems which have become highly privatised and market-oriented (Curry et al., 2012), a pattern also seen in the USA (Fuglie and Tool 2014). But characteristics of horticulture particularly stymie innovation, namely the power imbalances between producers and retailers, and large variations in business sizes and crop types which make it hard to agree research needs (Menary et al. 2019). This is not inevitable, as demonstrated by the health of horticultural R&D in the Netherlands – the country many UK growers look to jealously. The two had similar post-war trajectories, with declining public investment in R&D and increasing privatisation of extension from the 1970s. Pathways diverged by the late 1990s when the Dutch government became concerned about the risks of fragmentation, so merged state research institutes into a university, creating a specialist institution with the critical mass to lead R&D for future challenges (Spiertz and Kropff 2011). The UK lacks such a coordinated approach.

The third dimension of horticulture's skills crisis cuts across the previous two: the challenge of recruiting sufficient people. Data demonstrating the shape and scale of these shortages is inadequate. In 2018 the government began surveying horticultural businesses in England on their seasonal labour needs (Defra 2019a, 2019b). Results showed that between 25% and 36% of responding businesses needed seasonal labour, of which the proportion reporting a shortfall varied between 26% and 34%. Earlier data from an NFU survey suggested at least 4,000 unfilled vacancies (ONS 2019). Reporting initiated by the government to monitor pandemic impacts found 23% of growers unable to recruit sufficient labour to meet their productivity targets (AHDB 2020). All self-reported data should be treated with caution as it may be in businesses' interests to emphasise the economic threats of shortages. Also problematic is the lack of a standard definition of labour or skills shortages (Anderson et al. 2021). The most comprehensive survey found over half of businesses reported "hard to fill vacancies" mostly at managerial and supervisory grades, with almost half being seasonal roles (Pye Tait 2020). Respondents said it is difficult to find "semi-skilled workers who are reliable and interested in a career in the sector". Skills gaps within the existing workforce were seen to arise from new recruits lacking key skills and limited training once in work (ibid.); the skills crisis extends beyond recruitment into continuing professional development (CPD).

This characterisation of skills gaps exhibits the tendency to disregard informal learning and skills acquired on the job, as formal training is the assumed route to filling them (Pitt 2021b). It also shows the need to look beneath the surface of apparent skills shortages, as employers emphasise shortages of *quality* people:

> A common issue with respect to recruitment (rather than specific skills shortages) is that new workers lack quality, interest, and an appropriate attitude towards working in the sector.
>
> (Pye Tait 2020, 36)

A vacancy may be hard to fill because the employers' expectations are too high, rather than absence of available workers. As shown in Chapter 4, shortages are not simply a matter of numbers, as employers seek the right kind of employee: discourses around access to seasonal workers draw on socially constructed ideas of the workers horticulture needs, namely good migrant workers (Campbell 2019; Rogaly 2008; Scott 2013; Scott and O'Reilly 2022). But there are clearly problems with recruitment across horticulture, with shortages reported across specialisms and entry levels (Horticulture Select Committee 2023).

When I asked stakeholders about horticulture's supposed skills challenges, recruitment was most prominent. Not enough people want to work in horticulture:

> I think the issue largely, to me, is attracting people into the industries, that want to learn those skills, or regard it as a satisfactory career.
>
> (Sh9)

Recruitment problems were identified across the career pipeline, from school leavers, through entry-level roles and specialist senior positions, with several stakeholders emphasising that this applied to permanent and seasonal roles and predated Brexit. This challenge was seen to be at least a decade old, with some tracing it to urbanisation disconnecting people from land-based careers.

Growers often blame recruitment challenges on the sector's negative image and lack of public understanding of horticultural careers (Aldous et al. 2014; Horticultural Sector Committee 2023; Pye Tait 2020). Numerous initiatives have sought to communicate to young people the varied opportunities available, but their limited impact suggests more to the problem (Pitt 2019 and 2021b). Still, the idea that horticulture is unappealing persists with stakeholders telling me horticulture has an "image problem", associated with wider disinterest in agricultural work and ignorance about the food system. Others blamed horticulture, suggesting the term is not well understood, found "posh" or "unsexy"; horticulture's image associates it with hard, boring work and lower status careers. Interviewees mentioned individuals directed to horticulture because they struggled academically, or were regarded as naughty pupils, perpetuating the assumption that "No one with any ambition or intelligence would want to go and work in horticulture" (Sh12).

The roots of this image problem carry forth the way horticultural work was demeaned in the 19th century and suggest why it is so difficult to counter. One suggested factor was lack of exposure to people working in horticulture – not surprising given the distance between production and consumption, and deliberate isolation of workers. Horticulture was suggested to be associated with low status roles: "something your servants do" and "a manual labour thing" (Sh11). Stratification of jobs according to their supposed intellect has not served the sector well. But there are dimensions specific to horticulture and how it is regarded in relation to skill: "I think there's the assumption as well that everybody's a gardener" (Sh7). A nation of gardeners regards it as a self-taught hobby, not a profession:

> it becomes not seen as an art, it's not really a craft, it's not really seen as a science, it's just something that happens in people's gardens, and they don't really perceive it as a profession. There is a perception that it's a non-skilled career, by a lot of people, because they do it at home.
>
> (Sh11)

So horticulture is in a contradictory position of being so unfamiliar people overlook it as a profession, whilst suffering over familiarity from home gardening which positions it as an unskilled career.

The "image problem" is not wholly about lack of information or correcting mis-perceptions though, because the reality of the jobs can deter potential recruits:

> it's actually finding that labour, in any form, that is willing to do the tasks that are required in horticulture that are quite manual, quite back breaking, quite uncomfortable positions that people are in day in day out.
>
> (Sh9)

Another "elephant in the room" is pay:

> Money in the industry for a career can be rewarding, there are careers in the industry that you can earn substantial bucks. But the general, advertised levels are quite low.
>
> (Sh10)

Better pay and more interesting jobs were suggested routes out of the recruitment crisis, with some regarding innovations such as automation and controlled environment agriculture (CEA) as offering more appealing careers. But it will be difficult to reverse long-standing trends. Additional vacancies left by lower levels of EU migration are not appealing to the settled population (Nye and Lobley 2021).

Seasonal migrants' availability has shaped the nature of jobs dominating UK horticulture and employers' expectations. Their standards for "quality" workers influence how they recruit, and how the sector's "skills crisis" is constructed. Growers want employees with self-motivation, reliability and commitment, as suggested by a recent job description for production roles seeking a:

> team player, enthusiastic, reliable, flexible and adaptable demonstrating a 'can-do' attitude; excellent organisational skills and attention to detail; be passionate about what we do; an outstanding work-ethic and a pride in being part of the team!

Quite high expectations of a temporary recruit. Any employer wants good workers, but in horticulture, good work has been constructed around ethnic stereotypes. Albeit sometimes a little sheepishly, many growers described recruits from Eastern Europe in comparison to the domestic workforce as harder working, more reliable and better suited to physical work. I was struck by how growers commonly referred to each employee by nationality, like the trainer who told farm managers numerous anecdotes about motivating the workforce, each featuring "a Polish supervisor" or "a Romanian woman". He seemed to imply we expected them to be limited by their national origin, when in fact they are great workers. Eastern Europeans are regarded as good migrants who want to work hard to earn good money and so accept long hours of physically demanding tasks, whilst lazy locals rely on the welfare state or want softer jobs (Scott 2013; Scott and Rye 2021). Such stereotypes are not new, with successive generations dismissing today's youth as work-shy. They are also global, with migrants often described as better suited to horticultural work for their work ethic or physical build, drawing on highly racialised and gendered caricatures of minoritised groups (Freidberg 2009; Holmes 2013; Nieto 2014; Scott and Rye 2021).

The other problem with how employers construct worker shortages is that what they *say* makes a good worker is only part of the story. Characteristics associated with migrants include compliance, obedience and willingness to tolerate uncertain or even poor work conditions such as last-minute changes to work patterns (Rogaly 2008; Scott and Rye 2021). These are not stated in the job advert, but make migrant recruits desirable. Employers feeding globalised supply chains need flexibility, which for workers means precarity, and sometimes exploitation. During the pandemic, domestic recruits working alongside migrants brought this into relief. One grower told me local recruits spoke up for Romanians and Bulgarians who worked through breaks and sought overtime, saying this was exploitative, but their employer responded that the Eastern Europeans did not want rest to ensure they maximised their earnings. The employer told me this story to explain why seasonal migrants are more productive than locals; to me it showed that working up to or beyond the edges of fair and legal employment is expected of good workers.

What UK growers call their labour shortage is actually a lack of migrant workers and their particular qualities which make it possible for growers to operate flexibly. One recruiter described how in the good years production colleagues could request additional labour one or two days in advance, and intermediary agents would supply them. Now the lead-in time is weeks, or months if visas are required, assuming the agency can find recruits. These response times are too inflexible for just-in-time order fulfilment. For growers, "labour shortage" equals loss of flexibility and uncertain planning horizons; they lose the slack they used to have from over-recruiting to cover fluxes across the season. Employers might take on more workers than they need on an average day, knowing they can leave some in their accommodation on slow days. In the new just-enough employment market, short-term increases in workload can only be met through offering overtime, which those paid piece rates often accept. To employers, labour shortage might mean a lack of people to work above or below average weekly hours and can be code for loss of control over workers' employment patterns.

Lazy locals Pick for Britain

The 2020 season presented growers with a natural experiment testing their hypothesis that British workers are too lazy to do seasonal horticultural work, as they could not bring so many workers from overseas. The government supported a campaign reminiscent of wartime calls to "Dig for Britain", with citizens urged to "Pick for Britain". Huge publicity around the need for pickers backfired when the recruitment website instantly crashed. The scheme saw a low proportion of interested applicants ending up in jobs, and others complaining about them (Milbourne and Coulson 2021). Whether through Pick for Britain or direct recruitment, growers took on more domestic recruits than they had for 50 years, often people furloughed from other sectors. Whilst there were exceptions, growers reported lower productivity as the new recruits were slower and less effective. One grower calculates they need a worker for six weeks to pay back recruitment and induction costs; only 2 of 46 local recruits crossed this threshold. Once international travel restrictions were lifted, growers resumed reliance on migrants and furloughed workers returned to their old jobs. Many judged Pick for Britain a failure and used this to argue there is no local workforce to fill horticultural jobs (Milbourne and Coulson 2021). Some growers may have used the campaign cynically to demonstrate to the government the need to extend and expand the seasonal worker scheme. As one grower told me: "We now know definitively that the UK workforce can't do what these guys are doing".

Employers focused on the "lazy local workforce" revealed by Pick for Britain, but 2020 also revealed the jobs to be precarious and demanding (Scott and Rye 2021). These conditions were not new: migrants described a carrot processing plant as like a "cold hell", and piecework on strawberry farms so fast, with constant bending making injury inevitable (Rogaly and Qureshi 2017). But domestic recruits were perhaps more likely to judge them as unacceptable and better able to articulate concerns. This suggests that problems lie not with workers but the work (Milbourne and Coulson 2021; Scott and

O'Reilly 2022). But the tendency to blame workers recurs, so, for example, the welfare system needs changing to incentivise people to take up horticultural jobs (Campbell-Gibbons 2011). The parliamentary inquiries explored in the next chapter gave prominence to workers' deficiencies rather than systemic problems with food production and agri-food supply chains, or the jobs they create. The industry tends to see the image of agricultural work as a problem to be countered through PR and education, rather than through making jobs better (Devlin 2016; Pitt 2019). To be fair, little can be done to make tasks such as picking and weeding more exciting or less physical, and industry leaders *are* working to offer more enticing and rewarding work, through innovations which in any other sector would not be exceptional – family-friendly hours, year-round roles. There are some simply bad employers, as cases of exploitative practices highlight. But there is further reason to suspect that UK horticulture's recruitment crisis is not the fault of individuals who may or may not want its jobs, but rather due to the extent of bad work and agri-food systems which make it difficult for growers to offer better jobs. The fact is, there are considerable numbers of British people who *do* want to become food growers.

Outliers in the field

UK horticulture is bifurcated into extremes of scale, and this bifurcation extends to its recruitment patterns. Whilst large conventional growers bemoan the lack of interest from the domestic workforce, particularly younger generations, smaller growers and those operating to agro-ecological principles seem highly attractive. One interviewee who runs a horticultural business alongside working for an NGO promoting agro-ecological farming described the situation:

> It's got a training problem, and it's definitely got an access-to-land problem, and not-being-able-to-make-a-viable-business-because-the-price-of-food-is-very-low problem. But it doesn't have a recruitment problem.
> (Sh12)

He described a vibrant scene of new entrants wanting to work and train in growing, a majority of whom are graduates or career changers, many of them women. They aspire to grow food for ethical, political or lifestyle reasons, driving them to work in the alternative sector or establish their own enterprise (Soil Association 2018; Taherzadeh 2020; Styles et al. 2022). Larger companies which can promote themselves as sustainable, ethical businesses also thrive in the labour market (Pitt 2021b), suggesting growers who offer good work attract good workers. As the quote above indicates, just because there are good numbers wanting to work in agro-ecological horticulture, it is not crisis free. Multiple barriers make it difficult to train in sustainable horticulture, not least the lack of formal educational pathways, and reliance on entry via un- or low paid work experience (Taherzadeh 2020; Styles et al. 2022). As with any career development which requires independent means, this is exclusionary and means that people of colour are under-represented (Calliste et al. 2021; Weiler et al. 2016). The informality of these routes also leaves them open to exploitation and practices contravening labour law (Weiler et al. 2016).

Informal work arrangements make it even harder to quantify who works in agro-ecological growing, with many holdings too small to count in government statistics. The workforce in small-scale agro-ecological horticulture also includes significant voluntary labour (Ekers et al. 2016). There is no single body representing these growers who might join Landworkers Alliance, Organic Growers Alliance, Soil Association, Community Supported Agriculture (CSA) Network UK and Permaculture UK. Various platforms promote opportunities to volunteer on UK organic farms, including the worldwide network WWOOF; EU programmes such as European Solidarity Corps supported placements with horticultural social enterprises. A small survey of organic growers found just less than half of respondents pay everyone on their team, with 54% using some voluntary labour (Collyns 2022). This is a contentious issue, with differing opinions on whether it is right for food producing businesses to rely on unwaged workers (Chapters 6 and 7).

That people are willing to work in horticulture, doing what can be physically demanding and dull tasks without receiving pay, suggests agro-ecological growers have an appeal which large conventional growers lack – they struggle to recruit for *paid* roles doing similar work. So the claim that it is the work, not the workers, to blame for horticulture's labour shortages needs refinement: it's not the workers, it's the work *and* who the work is for. Values-driven employers who value their people and offer some control over work conditions can recruit. As will become apparent from the detail of life on smaller agro-ecological farms, the work can still be hard and under-rewarded, but these growers stick at it because they regard it as doing environmental and social good. It will also become apparent that economic viability shapes their knowledge ecosystem and employment conditions. It is systemic challenges which erode skilled work, not individuals in the workforce.

Conclusion: A history of devalued work and workers

This chapter profiled UK horticulture and uncovered roots of the workforce's devaluation. A perspective from one region showed the industry's gradual decline, and bifurcation into growers of very different scale and structure, shaped by global agri-food supply chains. The shape of the workforce is typical of this regime, with many temporary workers, largely immigrants. A historic perspective showed how farm work has typically been regarded as low status, deserving only low pay. Moments of crisis brought farmworkers into public and political consciousness, only temporarily and with little material benefit. We saw how narratives constructing this as unskilled work are long-standing, as industrialisation and urbanisation forged separation between rural and urban workers, shaping a public view of agricultural workers as lazy, uneducated and uncivilised. Spatial disjuncture reinforced pay disparities, which reinforced social devaluing of farmworkers: they earn low pay because they do unskilled work and must be uneducated if they receive such low pay. Perceptions of agricultural work as unskilled fuelled assumptions that those doing it must be unintelligent, so deserve only low pay.

Horticulture fared particularly badly in stratification of agricultural work based on judged levels of specialised skill. The more intelligent the animal the stockman (and it was nearly always a man) worked with, the more skilled and better paid the worker. Roles focused on plants did not feature in these notions of skilled work because constructions of work as skilled or not are highly caught up with who is most likely to do the work. Roles largely filled by non-professionals – women and children – or attracting migrants and marginalised groups such as gypsy travellers and the urban poor were most likely to be perceived as low or unskilled; workers' low social status reinforced this. This pattern is seen across nations and history, with more marginalised groups likely to dominate low paid roles, a situation justified by the construction of their work as unskilled. Horticulture's seasonal employment cycles heighten this because its reliance on temporary roles makes it easier to dismiss the workforce as unspecialised.

In Chapter 1, I proposed another reason horticultural workers have lingered at the bottom of the hierarchies of agricultural workers: plant blindness. In the hierarchy of beings common across Western thought, plants sit at the bottom. Until quite recently it has been common to consider plants as immobile, uncommunicative and lacking complex response mechanisms or anything akin to animal intelligence. Closer attention to their capabilities has revealed levels of complexity and ability which bear the hallmarks of clever beings. People assuming plants to be stupid would logically regard those who work with them as needing little education or skill to manage them. It is no great leap to suggest that underestimating what it takes to work with plants makes it easier to devalue those employed in horticulture.

Looking to the historic roots of the devaluation of horticultural work revealed narratives persistent today. The myth of the rural idyll has long included the idea that farm life used to be better, and this includes its workers. Employers have persistently rued the decline of hard-working agricultural workers and derided the current generation for their disinclination to work as hard. This narrative lives on, the contemporary version suggesting young people are more interested in having their hands on a phone than in the soil. Employers claim there is something wrong with this generation which means they won't do horticultural jobs, blaming them for a skills crisis instead of considering what might be wrong with the jobs on offer. When debating who should pick British crops, the problem is located with individual workers, as explored in the next chapter. Focusing on bad workers rather than bad work individualises challenges around skills rather than looking to systemic problems requiring collective responses. Countering shortages of workers is viewed as a matter of moving people with the right qualities to the right place, or convincing young people that they want horticultural jobs; in both cases the desired quality is associated with the individual. Horticulture's skills crisis is blamed on individual recruits who fail to understand the work, or are not interested in doing it, when the actual crisis is in the nature of the jobs and what makes them so unappealing. This crisis is rooted in the economic and political dynamics of the agri-food regime, which squeeze growers to invest as little as possible in workers and their skills. The true crisis for UK horticulture lies with the nature of our agri-food systems, but this is overlooked and individual growers are blamed for their own failure to recruit. The real crisis includes the injustice and inequality of racial capitalism which makes it possible for food growing in places like the UK to rely on seasonal migrants.

The same tendency to individualise the nature of skilled work and the challenges around it is apparent in narratives around another key dimension of the UK's horticultural workforce. In line with global trends, the sector came to rely on migrant workers, and this is entangled with what employers conceive as good workers. This reliance on migrant workers has skills implications and is shaped by problematic understandings of skill as a facet of an individual worker, dynamics which combine to underpin undervaluing of the work. Seasonal horticultural jobs dominated by migrant workers are more easily dismissed as unskilled because of the role's temporary nature: such short-term jobs cannot depend on high-skilled workers or allow skill acquisition. The dominance of non-British workers perpetuates the notion that the roles are unskilled, through racialised characterisations of the work as purely physical or manual, not requiring intelligence. These constructions essentialise the worker according to their ethnic background and essentialise their skill level as something they have or not – a fetishised commodity. The migrant *has* the desire and the ability, and the British youth *should* have it. And shortages of skills are seen to be a problem of too few individuals holding or developing horticultural skills, rather than problems with inadequate educational infrastructure, or lack of economic capacity to invest in professional development. Across narratives of horticultural work, skills problems are made a matter for individuals, with skill a free-floating capital which can attach to them or not. As Iskander argues, this is problematic for tending to demean any individuals regarded to not hold that capital (2021). To this I would add that it distracts from collective responsibility for ensuring essential knowledge is flowing; maintaining flourishing systems of horticultural knowledge is not something an individual can do, nor should they. A healthy ecosystem of horticultural knowledge is a common good.

4 Who picks for Britain?

Introduction

The coronavirus pandemic was a fertile period for those interested in horticultural work, as it gained unprecedented public attention. And just as growers faded into the background, Brexit and then Russia's invasion of Ukraine pushed their problems back into the spotlight. But a longer view highlights that what is portrayed as one short-term crisis after another, then another, is rather a system of reliance on temporary workers, mostly from overseas. Discourses around horticultural work and who should do it signal how immigration policies like the UK's construct workers and skill. As categories like race and gender have become unacceptable social markers, skill has stepped in to provide grounds to define and exclude people, with the advantage that it suggests categories based on merit, not biology (Warhurst et al. 2017). High skilled workers are welcome, low or no skilled ones are not. But notions of skill used to arbitrate immigration are imprecise; in the absence of a careful definition of skill, immigration rules refer to job specialisation or qualification. Categorisations are also influenced by colonialism, meaning they act as proxies for race (Liu-Farrer et al. 2021). Whilst the thinking behind skills-based immigration rules is fuzzy, implications for would-be migrants are clear. These policies tie skill to citizenship, with the unskilled denied social standing (Iskander 2021). As I highlighted in showing how horticultural work – like other "manual" jobs – became devalued, unskilled work is deemed such through assuming certain aptitudes as natural and associated with certain gendered, racialised bodies (Osterman et al. 2022). Tracking how horticultural workers have featured in these debates therefore reveals much about why their work remains under-valued.

This chapter traces these debates through media and policy discourse around horticultural work in the UK over recent decades. Media analysis focused on mainstream news coverage of horticultural work from 1990 to 2022.[1] Insight to policy debate focused on five parliamentary inquiries led by the Environment, Food and Rural Affairs (EFRA) committee during the same period.[2] Patterns of public attention to the topic highlight how attitudes have evolved, and moments affecting the sector. Media narratives also demonstrate

DOI: 10.4324/9781003281689-4

how horticultural workers have been perceived, signalling how different stakeholders construct the work around their own interests.

Looking back to 1990, for considerable periods, horticultural work received very little attention. Figure 4.1 indicates how many news stories referred to horticultural work annually, showing an upward trend through the 2000s, peaking in 2018. Peaks in attention coincided with events or policy changes impacting labour availability, prompting farming organisations to publicise growers' concerns (Tables 4.1 and 4.2). A gradual turn of attention to horticultural workers across the 2000s reflects heightened concern with immigration, culminating in the vote to end the free movement of workers from Europe through exiting the EU. In this ongoing narrative – disrupted by the pandemic – horticulture is associated with low skilled migration of the kind opposed by anti-immigration voices. Migration politics have been central to attention to horticultural workers.

Outside broader immigration debates, horticulture has been largely disregarded by the media, allowing the general public to ignore the people who pick their fruit and veg (Scott and O'Reilly 2022). Workers only become visible at times of crisis, when they appear in the form of potential shortages, with implications for British farmers or food supply. Ironically, horticultural workers are most apparent as an absence. And when they are present in the media, workers are discussed as numbers – the 000s of visas, or the tens of thousands of unfilled vacancies – not as people. This is symptomatic of a tendency for migration discourse to focus on numbers per visa category, not actual people (Liu-Farrer et al. 2021). Individuals doing the work are rarely present, as the concern is their labour and its productive value, not their lives; seasonal workers are labour force rather than full persons (Hahamovitch 2003; Iskander 2021). Exceptions are occasional profiles of workers characterised as either

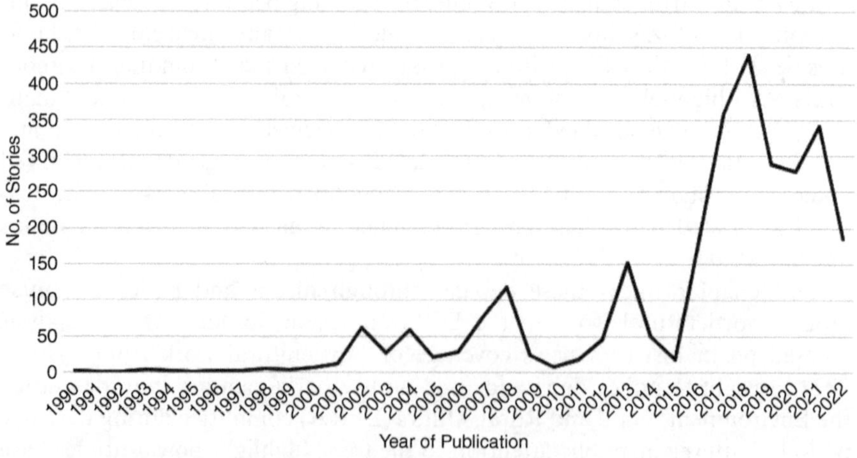

Figure 4.1 Volume of UK print media coverage of horticultural work since 1990.

Table 4.1 Key moments reflected in media coverage of horticultural work since 1990

Year	*Events influencing coverage of horticultural work*
2002	Labour Government proposes more routes for migrant workers including a broader seasonal worker scheme.
2003	Operation Shark uncovers malpractice by gangmasters including exploitation of seasonal agricultural workers
2004	Freeze in visas for seasonal farm workers following whistle-blower in Romanian Embassy
	Dispute over planning permission for accommodation for strawberry farm workers in Herefordshire
2007–2008	Concern over reduced SAWS quotas
2012–2013	Concern over the end of SAWS
2016–2017	Brexit vote prompts concern over decreased levels of EU migration
2018–2019	Seasonal Worker Scheme pilot announced, followed by problems with implementation
2020	Covid-19 pandemic reduces the availability of seasonal migrant workers; national campaigns to recruit domestic workforce
2021	Labour shortages across the economy; food supply chain issues peak at Christmas; calls to extend seasonal visas to other sectors
2022	Russia's invasion of Ukraine exacerbates labour shortages
	Investigations of exploitation of workers on seasonal worker visas

Table 4.2 Main narratives of news stories on horticultural work 2000–2022

Narrative	*No. of stories*
Seasonal workers schemes: changes to policy, general discussion of migrant seasonal workers	25
Reporting current or predicted farm labour shortages	25
Concerns regarding Brexit's impacts	16
Covid-related impacts	15
Making the case for seasonal worker visas or scheme	17
Broader labour crisis including food growing	13
Immigration figures and trends with mention of seasonal agricultural workers	13
Pressure to reduce immigration including seasonal migration	9
Calls for the domestic population to do seasonal work	6
Feed the Nation & Pick for Britain campaigns	6
Nature of seasonal work – negative accounts	6
Nature of seasonal work – neutral accounts	6
Impacts of invasion of Ukraine, including on Ukrainian seasonal workers	4
Why the domestic population won't do seasonal work	3
Technology/robots ready to replace workers	3
Brexit: positive impacts	2
Gangmaster problems and illegality	2
Industry working to attract workers	2

"hardworking hero" or "exploited victim". The absence of workers' voices leaves the majority of opinions coming from employers and their representatives, such as farm unions and grower groups, with politicians often voicing opposing views. These dominant voices construct the social imaginary of horticultural workers, seasonal workers in particular; we should therefore be mindful of who benefits when workers are imagined as low skilled, or when shortages are claimed (Campbell 2019; Rogaly 2008). The government is not always convinced by growers' narratives, as apparent during parliamentary inquiries. But horticultural businesses have been broadly successful in convincing policy makers that continued access to seasonal migrant workers is preferable to alternative scenarios which would see crops wasted, or produced overseas (Scott 2022). Media analysis suggests six key phases or moments, now explored in turn.

The dirty secrets of migrant work, 1990–2010

During this quietest phase of discourse, narratives can be summarised by two words: ignorance and illegality. There was little public attention because growers had access to the workers they needed until the silence was broken by cases of illegal employment and concern with illegal immigration. Horticultural workers appeared as something of a dirty secret for consumers unaware of their significant role in food growing, or of the exploitation this could represent. It was not in employers' interest to reveal their secret, until attitudes towards immigration began to shift, and their labour supply was threatened. Around 2002 Labour Home Secretary David Blunkett prioritised reducing illegal immigration by expanding seasonal worker programmes as a legal route to work (Blunkett to allow in more foreign labour, The Mirror, May 29, 2002).

Awareness of considerable numbers of Eastern Europeans working in horticulture was increasing, often associated with problematic gangmastering. Gang labour has been used in UK agriculture since the 19th century when agents or gangmasters contracted teams of temporary casual labour. It became prominent as neoliberal governments shaped flexible labour markets from the 1980s (Strauss 2013). Horticulture gained attention for the scale of gangmastering, as in this story:

> It is estimated that at least 2,000 gangmasters are operating within Britain and that they are supplying as many as 75,000 workers a year to farms and packing houses.
>
> The illegal gangmasters have given the trade its dubious image, though there are many who operate within the law. The criminal gang leaders move around farms according to need and there is rarely a paper trail. Many jobs are organised by word of mouth and payments are always cash in hand.

> Farmers and growers pay the agent for the labour and are spared the expense of paying and administering tax and national insurance contributions. The workers, usually asylum-seekers without a work permit, or benefits cheats, also include white British nationals who like to moonlight and dodge taxes.
>
> (Gangmasters who reap the human harvest, The Times, July 8, 2003)

Although not necessarily illegal, and long dissociated from tied labour, gangmastering remained associated with unfree labour and enabled flexible labour markets (Strauss 2013). Criticism intensified in response to the deaths in 2004 of Chinese cockle pickers illegally employed by gangmasters in Morecambe Bay, and subsequent Parliamentary work to improve legislative controls.

Ground began to shift through the mid-2000s as the government was pressed to limit legal migration from newly acceded EU states, and pending members Bulgaria and Romania. Coverage suggests a concern with illegal migration evolved into unease with low or unskilled migration. Horticultural workers were portrayed as a separate category of migrants, only staying for a season. But anti-migration voices such as Migration Watch opposed expansion of SAWS, describing it as a slippery slope towards lax controls, or an open door to overstaying. Right-wing concern with immigration was fuelled by an embassy whistle-blower who highlighted irregularities with seasonal worker visas, prompting the closure of SAWS in 2004:

> Growers, who have long-standing links with colleges in Bulgaria and Romania, are furious after this year's student workforce was signed up as long ago as last autumn. Each person has had their details processed and a work permit is ready for them under the seasonal agricultural workers scheme, but they are unable to obtain visas to travel from the British embassies in Sofia and Bucharest.
>
> (Fruit growers hit by visa ban on migrant workers, The Times, May 17, 2004)

This article reflects this period's discourse around seasonal workers. Firstly, they are referred to as students on working holidays suggesting a benign form of migration: educated people who will not settle. Secondly, to resist anti-immigration arguments, the farming lobby portrayed migrants as an essential and established resource: farmers have relied on them since the war. Finally, the story foregrounds negative impacts on growers, ignoring the people expecting to come to the UK.

Through the 2000s whispers of the presence and significance of seasonal workers were heard in longer profile pieces. Read today these convey a sense of revealing the secret of these workers, and surprise at their presence in rural England; who would imagine Brazilians working on Kent's hop harvest

(Hopping to a samba beat, The Times, September 14, 2002). Profiles of seasonal workers tended to one of two types: exposés of the tough, even exploitative work and conditions, such as the Guardian's campaign to reveal labour in the food chain (The new landless labourers, Saturday 17 May 2003) and similar:

> Revealed: Scandal of Britain's fruit-farm workers; These Bulgarians are flown to Britain, live in packed caravan compounds and pocket just £45 a week to pick fruit for Britain's biggest retailers.
>
> (Exclusive Tesco promises to investigate working practices of fruit supplier, The Independent, July 10, 2009)

Alternatively, the veil was lifted to reveal hardy migrants doing work that British workers would not. Weather often featured in the challenging settings these heroes battle through:

> The drizzle was incessant and the daffodil fields were shrouded in early morning mist. But Kalina Nikoleva was undeterred and even managed a smile as she stood knee deep in mud [...] Wrapped up in sweaters, and wearing waterproofs, boots and a woolly hat, Miss Nikoleva was fully protected from the miserable weather. She had plasters on her fingers and surgical gloves to protect her hands. "I am having fun, but it is hard work," she said.
>
> (Would you pick 8,000 daffs for pounds 50? Kalina does, The Daily Telegraph, January 6, 2007)

This story goes on to describe her family back home, her plans to spend the season's earnings on starting a business, then details how SAWS works. A representative of the agency which placed Kalina described Eastern Europeans on the scheme as "very, very hard working [...] But whatever the weather, they don't complain. They want to earn their money and get their lives started at home". The "tough but rewarding" narrative sometimes comes directly from the workers; it is also deployed by employers and recruiters to argue Eastern Europeans *want* to do the work (Rye and Scott 2021). This became increasingly important towards the decade's end as SAWS shrank, then ceased, and growers expressed concern for labour shortages. Pro-SAWS voices distinguished it from other migration, as then chair of the NFU's Horticulture Board argued:

> The SAWS scheme has run efficiently for decades and the vast, vast majority of those who take part simply return home to carry on with their studies. The Home Office somehow seems to think this is an issue about mass migration but that is simply not the case. We face the risk of growers simply deciding to give up and grow other things.
>
> (Strawberry fields forsaken, The Independent, May 12, 2008)

But by now the Labour government was tougher on immigration, so they positioned farm work within unskilled migration, closing SAWS in 2013 as part of "wider policy of cutting down on non-EU migrants performing unskilled jobs" (Strawberry fields forsaken, The Independent, May 12, 2008).

Hostility hardens to hard-working Eastern Europeans, 2010–2016

At the time no one knew this was the pre-Brexit era, but in retrospect, indicators of sentiments against free movement are clear. A 2010 editorial signals attitudes turning against "unskilled immigration":

> The Chinese have picked the asparagus, the Pakistanis have wiped down the operating theatres, the Afghans provide minicabs, the Nigerians have driven the Tubes, the Australians have pulled the pints and the Poles have dug the foundations of our new schools. Only two years ago £13 million of British fruit went unpicked because farmers couldn't find enough British labour to harvest their crops, forcing the Government to raise the quota for migrants under the seasonal agricultural workers' scheme. But the rest of the country has suffered.
>
> (Thanks, immigrants, but times have changed, The Times, June 9, 2010)

A new government, the austerity era of reduced public investment and perceptions that too many people from new EU states had settled fed anti-migration sentiments and a "hostile environment" immigration policy. Yes, the economy had made good use of workers from around the world, but for many this did not outweigh negative impacts.

Between the cessation of SAWS and the Brexit vote, coverage was dominated by growers' calls to reinstate seasonal visas. Pleas to address labour shortages appealed to the emotive image of rotting fruit and vegetables (Curb on migrant pickers 'will condemn crops to rot', The Times, September 13, 2013), or likely impacts on consumers (Food prices to rise 15pc 'unless ministers let in more migrants', The Daily Telegraph, May 15, 2013). The government resisted, arguing economically inactive Britons could fill the vacancies (Seasonal migrant workers scheme closes to help British into work, Guardian, September 12, 2013). Divergent opinions on the potential for a domestic horticultural workforce are prominent throughout this period, entrenching a long-standing disagreement. Growers argued the domestic workforce won't do seasonal horticultural work: it is not economically attractive enough to those on benefits (Romanians will soon be free to do other jobs, Daily Mail, May 15, 2013). Those who try it "can't keep up with Polish workers", or don't stay the course (Pick fruit and veg or lose your benefits, minister tells Jobless, The Times, February 26, 2014). Conversely, the government argued British citizens *could* take the jobs but are too lazy or choosy, with one farming minister finding it unacceptable "that people who could work don't take work because they think there is something unglamorous about farm work" (ibid.). Conservative MPs representing constituencies with considerable horticultural industries challenged their government's line (Seasonal migrant workers scheme closes to help British into work, Guardian, September 12, 2013). Their minority position indicates how the issue challenges the Conservative Party's traditional allegiances, pitting the needs of rural business against anti-immigration supporters (Scott 2022). These tensions resurfaced across the next decade, as the farm lobby made an ever-louder case for more immigration, supported by rural Conservative MPs arguing seasonal workers need a welcome, not a hostile environment.

Brexit Looms, 2016–2019

In 2016, a narrow majority voted for the UK to exit the EU, bringing uncertainty regarding the movement of workers and products outside the European common market. Amongst a fraught, often hyperbolic debate, horticultural work was aligned with un- and low-skilled work, as when Brexit supporting Conservative peer Lord Lawson made controversial comments on skills-based immigration:

> it's people like the Romanian fruit pickers that we don't need. It is not the highly skilled people that the City wants – who will always be welcome.
>
> (Brexit: Bankers are welcome in UK but 'we don't need Romanian fruit pickers', says Lord Lawson, The Independent, April 28, 2017)

Whilst party colleagues distanced themselves from his remark, targets for reducing immigration pressured sectors like horticulture. In response, the industry portrayed farm work as deserving special treatment. As a farmer featured on the BBC put it:

> This isn't anything to do with migration or immigration, this is just seasonal workers coming over here, working hard, benefiting our economy and then going home.

But right-wing press interpreted this as anti-Brexit propaganda ('Cropaganda! BBC is accused of anti-Brexit agenda [...]', Daily Mail, March 21, 2017). That a farmer featured on Countryfile – a gentle Sunday afternoon showcase of rural life – prompted such backlash indicates the toxicity of debate.

Attention to horticulture increased significantly during this period, as the sector highlighted labour shortages post-SAWS, and as potential migrants were deterred by post-Brexit uncertainty and rhetoric. A recruiter I spoke to in 2018 described it as one of the most stressful times in her 20 years in the industry. The sector lobbied hard for new seasonal work visas, appealing to the need to protect domestic food production; the head of one of the country's largest growers, G's Fresh, argued:

> The government has to make a decision: either we bring the people to the work or we take the work to the people," he told the Guardian. "The government has to decide does it want [the UK] to produce food or not – that is their decision.
>
> (Brexit could herald end to British fruit and veg sales, producers warn, The Guardian, August 3, 2016)

In the same article, a soft fruit grower posed the rhetorical question: "Do you want all your fresh produce to come from foreign countries?", appealing to patriotic tendencies of the ruling Conservatives, and anyone concerned with food security.

Appeals intensified, in both volumes of coverage and increasingly emotive language. The sector claimed to face not just labour shortages, but a "labour crisis", which combined with other pressures to shape "cataclysmic" prospects for the industry. To bring these threats closer to home, potential price rises were cited, including an uplift of 50% for strawberries, highlighted by British Summer Fruits in the run up to Wimbledon. Wimbledon wouldn't be the same without British strawberries, and there are no British strawberries without Eastern European workers. The distasteful, even immoral, vision of fruit and vegetables "rotting in the fields" without people to pick them recurs through 2018–2019. Industry representatives referred to thousands of tonnes of produce wasted in 2017 due to labour shortages, calling for action to avoid similar wastage. The specter of 000s of punnets of soft fruit, or 000s of pounds worth of broccoli unpicked, then rotting, was conjured to convey why tackling horticulture's looming crisis was both moral and economic imperative.

Growers continued to argue the need for migrant labour because Britons won't do these jobs ('We've had one British fruit picker on the farm in five years. He quit after a day', The Times, June 23, 2017). Whilst the right wing were not all prepared to dismiss a generation of lazy British people, they perpetuated the image of striving Eastern Europeans "perfectly prepared to slog all day, in wind and rain and cloying mud". From their patriotic perspective, migrants had fled to better opportunities and pay in the UK (A Rotten Business, Daily Mail, October 21, 2019).

Tensions within the Conservative party intensified, as actors such as The Country Land and Business Owners Association, "traditionally a bulwark of Tory support", voiced the need for a different immigration approach (Warning over rural labour shortage, The Times, June 8, 2017). The farm lobby and DEFRA won concessions from the Home Office and the anti-migration lobby, with a pilot seasonal workers visa for 2019–2020. Its limited scale and time frame were immediately criticised by growers as insufficient (Brexit: farmers criticise temporary agricultural worker visa scheme, The Guardian, September 6, 2018). The 2,500 visas compared with 20,000+ under the previous scheme, and the 70,000 farm unions wanted, was portrayed by government as "a tightly controlled regime" and transitional measure (Visas for fruit pickers, Daily Mail, September 6, 2018). What it should transition towards was unclear, with then DEFRA minister appealing for more "creative" solutions in future (Scottish farmers' call to employ EU workers after 2019 wins Gove backing, The Daily Telegraph, February 21, 2018).

Feeding the nation?

Industry's concern regarding labour led the relevant parliamentary committee to hold an enquiry, to verify whether agriculture and horticulture were suffering shortages (EFRA 2017). They found it to be an established problem, intensified by Brexit which made potential migrants feel unwelcome. Broader trends, such as the changing value of sterling and improving living standards in Eastern Europe, combined to make work in the UK less appealing (EFRA 2017). The government disputed the shortages, tabling alternative migration figures and labour need, but ultimately the committee sided with growers, criticising the government's inadequate data.

Each side of the disagreement constructed the scale of labour shortages to achieve their preferred outcome – more migrant workers or fewer. Growers emphasised the need for seasonal migrants, arguing this as a matter purely of labour access. No doubt conscious that the Brexit vote indicated public dissatisfaction with immigration, they sought to distinguish their workforce from other migrants (Scott 2022). Needs, together with seasonal migrants' limited entitlements, were used to justify claims that "seasonal work is not a migration issue" (Hugh Lowe Farms, Written Evidence 2017).[3] Growers offered up seasonal workers rights in return for the right to work:

> We must ensure that they do not receive unlimited access to our welfare benefits from day one and they are only permitted to settle with their families after a number of years.
>
> (Richard Rowan, Written evidence 2017)

Limiting seasonal workers' citizenship makes them into the perfect migrant who gives work and receives little in return (Hahamovitch 2003). SAWS participants were referred to as "mostly students", making them quality, low risk migrants. Stakeholders also argued horticulture's special case on the grounds these jobs struggle to compete in an open recruitment market:

> Picking Cabbages, in December, in darkest Lincolnshire at -4 with sideways sleet is somewhat less attractive than being a barista at Pret or Costa.
>
> (HOPS Written Evidence 2017)

Immigration rules which tie workers to agricultural roles are preferable, as growers want labour as predictable capacity, not as people with complex lives.

Horticulture's labour shortages were blamed on short- to medium-term factors, Brexit in particular, with only passing indications of deeper problems. Issues with the broader food system only featured when producers fended off appeals to increase pay by positioning themselves as vulnerable to retailers:

> Farm businesses are price takers rather than price makers. If employers in the sector are forced to pay more for labour – which can account for in the region of 50% of a farm's production costs – somebody has to pick up the bill, whether that is further up the food supply chain or the consumer; and this is unlikely given the balance of power in the supply chain.
>
> (NFU Supplementary evidence 2017)

Minority voices situated labour problems within wider pressures on employment practices:

> The current agricultural market has created a race to the bottom on labour costs and worker conditions that are increasingly insecure, particularly in the horticultural sector. Looking forward, our vision should be not to replicate this model, but to build an agricultural policy that is more sustainable, both environmentally and socially.
>
> (FWAG Written Evidence 2017)

These contributions discussed jobs not just quantitatively – how many vacancies – but as qualitative concerns:

> at issue is not just a capacity question of who will grow our food, but an ethical question of how they will be treated.
>
> (Food Ethics Council Written Evidence 2017)

These voices were not blaming producers for offering "bad" jobs, or relying on seasonal migrants, but recognised growers as subject to systemic forces such as EU agricultural policy, reducing the potential for good jobs. From a trade union perspective, the problem was inequality:

> Workers deserve a bigger share of the wealth generated by this industry. This applies as much at the field harvesting end of the supply chain, as at the supermarket shop floor. […] To make food and farming a sector that is attractive to workers means a massive shift in attitudes to workers and to their pay.
>
> (Unite Written Evidence 2017)

Unite's proposed solution was for producers to push harder for higher prices; others sought attitudinal change:

> We need a new approach, based on attracting new entrants to farming, and promoting a new aspirational message that farming is a skilled profession premised on supplying people with healthy food whilst enriching the environment.
>
> (Shared Assets Written Evidence 2017)

In fact, these solutions are connected:

> The problem, it appears, is not that farm work in general lacks the requisite lure for British workers, but that the current system is tailor made to put them off.
>
> (FWAG Written Evidence 2017)

These minority perspectives positioned growers as victims of food systems in which they lack power to make prices and as culprits making workers suffer. But systemic pressures and solutions were overlooked by the committee, which did not address longer term challenges in their recommendations which focused on the immediate crisis and solution: more seasonal migrant workers.

(Y)our country needs you, 2019–2020

Had there been no Covid-19 pandemic, the UK government might have forged ahead to reduce low skilled immigration, urging sectors like agriculture to prioritise "investment in staff retention, productivity, technology and automation" (Immigration reforms 'will stop firms using cheap foreign labour', The Times, February 19, 2020). Instead, the pressures horticulture already complained of intensified, as social and economic restrictions immediately impacted the food system, particularly fresh produce supply chains (Garnett et al. 2020; Richards and Rickard 2020; Sanderson Bellamy et al. 2021). The timing of international movement restrictions was significant, as they commenced just when the year's first seasonal workers should have arrived. By March 2020 farming organisations called for emergency measures to ensure sufficient workers to plant and pick crops (Farmers call for 'land army' to sustain UK food production during coronavirus crisis, The Guardian, March 20, 2020). The sector appealed to its special status for exemption from limits to worker movements and chartered flights to bring them from Eastern Europe. Outlets typically opposed to immigration described this as a move to "help farmers" (Romanian Fruit Pickers Flown in to help farmers short of labour, Daily Mail, April 16, 2020). The need for essential workers trumped opposition to immigration.

The crisis was deemed worthy of a wartime response, with calls for a "land army" deliberately echoing World War II rhetoric. Campaigns appealing to the domestic population to "Pick for Britain" played on nationalism and patriotism to "Feed the Nation" (Scott and O'Reilly 2022). Campaigns did not shy away from how tough the work is, with one TV ad featuring tired limbs resting in a bath, close-ups of dirty hands and feet, faces battered by rain (Adam & Eve n.d.). Harvest workers featured as heroes battling through challenging conditions, rewarded by good pay and the knowledge they are feeding others. The narrative echoes that from the early coverage of Eastern European pickers who found the job tough but worth it. Imagery of a striking woman looking into the distance recalls depression era farm labourers captured by Dorothea

Lange; the dirty faced man with sweat in his hair echoes portraits of first responders to 9/11. These are people doing important work, in the toughest of conditions.

The creatives behind the campaigns were clearly good at their job, as an estimated 160,000 people responded, immediately overwhelming the website (Pick for Britain website crashes moments after minister urges nation to sign up, Mirror, 19 May 2020; Adkins 2020). But the nationalistic impulse to celebrate heroic domestic horticultural workers was soon undermined, as most who registered interest were never placed on a farm and they eventually represented only 11% of the season's workforce (Brits only small part of harvest workforce, The Western Mail, November 4, 2020). They were unfavourably compared with migrants:

> Every ton of asparagus not picked costs a grower about £1,700", he [Per Hogberg, of the Asparagus Growers' Association] said, and the slower pace of new recruits compared with returnees costs a farmer 15-20p per kilo picked. "It's backbreaking work," he added. "Migrant workers will go a 12-hour shift. Are UK student labourers or furloughed staff willing to put that in work? I don't think so".
>
> One asparagus grower in the east of England said that he expected to ditch 30 per cent of his crop. The farmer, who asked to remain anonymous, said: "If I had to rely on British workers I'd be out of business next year. It's an uncertain time at the moment but the government is somewhat deluded in thinking the 'great British workforce' is capable of doing this. They find it too physical. The English workers have been picking about 60, 65 per cent of what my returnees do. Out of 50-odd workers I've got about four good ones".
>
> (Farmers fear disaster as new Land Army enters the field, The Times, May 2, 2020)

The domestic workforce failed growers during 2020, but being forced to test their potential served employers' longer-term needs by providing evidence that British workers can't do these jobs. Some British citizens reported trying to secure jobs and being discouraged ('Just not true' we're too lazy for farm work, say frustrated UK applicants, The Guardian, April 20, 2020), but the dominant narrative promoted migrants as essential.

As Scott and O'Reilly argue (2022), narratives across the Pick for Britain campaign were managed to laud the hero migrant at the expense of the reluctant Brits, with heroic work through tough conditions celebrated, rather than highlighting the need for better jobs. Britons were presented as personally deficient to distract from structural failings preventing decent work, with most sympathy going to the victim farmers. This "selective and managed unveiling of precarious work" left commodity fetishism and the rural idyll in place (Scott and O'Reilly 2022, 11). Coverage largely favourable to employers supported their effort to construct labour shortages as resolvable only by flexible migrant workers. The farming lobby seized on this to reinvigorate their campaign to expand the seasonal worker scheme.

The crisis that ruined Christmas, 2021–2022

Just when growers thought things couldn't get any worse, 2021 brought widespread workforce shortages. Concern built through the year, as more industries and iconic companies reported issues, peaking with panic around Christmas. The vision of a celebratory family dinner missing British turkey, sprouts or spuds was painted to highlight failings in post-Brexit immigration rules, or the need for emergency support to recover from the pandemic. Depending on political allegiances, media sources used a lengthening list of shortages of products and workers to bash an incompetent pro-Brexit government, or industries "addicted" to cheap migrant labour.

In response horticultural actors promoted their usual solution: more seasonal labour. Food shortages helped make their case, with powerful images of empty supermarket shelves deployed in political tussles; rotting fruit and veg was joined by empty shelves and plates to tug politicians' heart strings. But the government was again locked in dispute (Ministers in immigration row over raising cap on seasonal farm labour, The Times, April 22, 2021). The Home Office resisted DEFRA's request for more seasonal visas, as key Conservative support bases – rural business, anti-migrants – voiced opposing needs (Scott 2022). On both sides the pandemic's power to reveal and celebrate essential horticultural workers had not endured.

Growers accessed seasonal workers through a visa scheme expanded for 2021–2023, which they argued was inadequate and poorly executed, causing continued labour shortages. They appealed for more seasonal migrant workers, justified by the failure of Pick for Britain. As other parts of the food supply chain such as haulage and meat processing struggled to recruit, they called for similar treatment, but the government was unsympathetic:

> A government spokesperson said: "The British people repeatedly voted to end free movement and take back control of our immigration system. Employers should invest in our domestic workforce instead of relying on labour from abroad".
>
> ('The anxiety is off the scale': UK farm sector worried by labour shortages, The Guardian, August 25, 2021)

Right-wing press blamed the industry, and located the solutions with them – pay pickers more, even if food prices rise as a result:

> harsh though it may seem, we should surely temper our addiction to cheap food – can it be right that a whole supermarket chicken costs £3? – if it means better wages for British workers.
>
> And if it is a shortage of butchery skills which threatens the existence of tens of thousands of pigs, then the big private abattoirs and butchers

> should have thought ahead, recruiting, reskilling and paying appropriate wages.
>
> (How rich for fat cats to squeal now they can't rely on addiction to cheap labour, MailOnline, October 5, 2021)

Although farming ministers were slightly more sympathetic, the government said food producers could not expect to be repeatedly bailed out; seasonal migration should be interim. Conservatives held to their vision of a post-Brexit high wage, high skill economy, with new immigration rules designed to reduce reliance on "cheap foreign labour" (Immigration reforms 'will stop firms using cheap foreign labour' The Times, February 19, 2020).

The government's logic was that 2021's labour shortages be borne as "temporary pain in the service of long-term gain" to deliver Brexit and help "wean the UK off low-skilled immigrant labour that once came via EU freedom of movement" (Britain is learning the hard way that migration can't be turned on or off like a tap, The Guardian (London), October 12, 2021). Critics suggested this failed to understand migration or economies:

> The UK's post-Brexit immigration system has been sold to us as one in which the country can cherry pick the world's "brightest and best" (a euphemism for the rich and highly educated), and dispense with the services of the less affluent. In reality, immigration will continue to play a significant role in sustaining those occupations deemed "low skilled", but which in fact are essential to the functioning of the country.
>
> (MailOnline, October 5, 2021)

Those challenging businesses "addicted" to cheap migrant labour thought they could attract workers to vacancies, seizing on the example of a broccoli picker paid £30 an hour to signal that growers *can* pay more (BIG BUCKS for blue collar workforce!, MailOnline, September 24, 2021). Such reports implied the public should be incredulous that such a low skilled worker could earn as much as more qualified professionals (Broccoli pickers paid £30 an hour as Britain runs short of vegetables, The Times, October 4, 2021). But those close to growers suggested that as "price takers" with "little power within the wider supply chain", they are unable to raise wages (AHDB Written Evidence 2017).

Skills mean points, points mean prizes

As the UK's new immigration rules were defined, growers identified the lack of routes for lower-skilled immigrants as "the biggest labour threat to the UK food industry and consequently UK food security and resilience" (ALP Written Evidence 2020). The EFRA Committee returned to the matter of labour, this time considering the wider food supply chain with a focus on immigration policy. Lines of debate were familiar from 2017, with government and industry still disagreeing on the scale of labour shortages and seasonal worker requirements; if anything, mistrust had worsened. The problem with points-based immigration was that it caters to medium to high skilled permanent roles, whilst seasonal work is temporary and "despite being high value in terms of importance to the nation's food security, as well as other wider public goods", is classified as low skilled (NFU Supplementary Written Evidence 2020).[4] There were signs of more nuanced thinking around domestic recruitment, but growers favoured the same old solution:

> We require a transient labour force who are not looking for permanent residency status.
>
> (Brassica Growers Association Written Evidence 2020)

They wanted a bigger, better, longer seasonal workers scheme, and the committee tended to agree, recommending a "generous cap" on the number of seasonal worker visas.

What is notable about this debate is how, for the first time, industry and politicians really considered the work done by horticultural workers, and whether it is skilled. An immigration system designed to make the UK a high-skill, high-pay economy incentivised employers to portray their employees as skilled; skill could now unlock doors to migration, and horticulture wanted that key. Stakeholders adopted multiple skills focused strategies to argue for more migrants: a small number suggested that low skilled work is equally essential to the economy, so it should not be excluded from immigration. One producer highlighted lessons from the pandemic that "'vital skills' cannot simply be defined by two measures: salary level and formal qualifications". They drew parallels with other "heroes":

> without the work of hospital cleaners the NHS would not have been able to respond, as it has done, to the unprecedented demands put upon it during the Covid-19 crisis. Under the new points-based system a person applying for a job requiring such 'vital skills' would not be admitted to the UK. For each element of the food supply chain there will be parallel examples of where a particular skill set will be a vital ingredient in enabling the whole chain to function effectively.
>
> (G's Fresh Written Evidence 2020)

G's Fresh went on to pursue the second strategy: challenging how immigration policy defines skill by educational qualification. For roles in food supply:

> However, such skills are often acquired through on-the-job training and experience but will not ultimately be recognised by either the payment of a high salary or the awarding of a formal professional or academic qualification.
>
> (G's Fresh Written Evidence 2020)

The NFU argued a range of agricultural jobs including produce packers and farm labourers are skilled on the grounds of requiring "excellent eye hand co-ordination, stamina, physical fitness and dual manual dexterity". Examples from horticulture included those able to "cut cauliflowers with a very sharp knife" (Cornwall Council Written Evidence 2021), and mushroom harvesters whose work is:

> complex, can be difficult to master, and takes approximately 12 weeks to learn. It has to be performed with a considerable degree of care and application of good judgement.
>
> (G's Fresh Written Evidence 2021)

The sector argued its efficiency depends on these skills, so people should be allowed to enter the country to work and upskill. To evidence the skill of their workers, growers highlighted the difference it makes to deploy:

> reliable workers who come back every season, who are already trained, for whom it is 'their' job and with whom they develop a good relationship.
>
> (ALP Written Evidence 2020)

Returners are up to 20% more productive than novices (GLFB Written Evidence 2020) because they have developed skills on the job. The committee agreed the policy's definition of skill was overly narrow and called for a more practical definition acknowledging the reality of food-related jobs:

> people are doing things that are difficult, that take time to learn, that have to be done with a degree of care and attention but that are not currently defined as skilled.
>
> (EFRA 2017, p16)

They appreciated roles classed as unskilled entail: "a level of tacit knowledge, gained through experience in the industry" (EFRA 2017). But skill of the role was emphasised over skill of the migrant, to make the case for 'entry level' routes for migrants who don't already have essential skills (FDF Written Evidence 2020).

Other arguments traversed the low-high skill categorisation to seek favourable positions for horticulture, for example, highlighting that the industry encompasses low and high skill roles. Some actively sought to challenge such categorisation, tending more to the argument underlying my definition of horticultural skills. The NFU identified what they called "an intellectual bias" in the government's skills definition as it valorised academic over vocational skills, neglecting learning through practical experience (NFU Written Evidence 2020). They challenged the term "low skilled" as "demonstrably derogatory to workers" and called for the government to cease using it. Others questioned the validity of terminology by placing low, high or unskilled in scare quotes, suggesting they recognised the social construction of skill categories.

The final ground for challenging how the new immigration system affected horticulture was the equation of low skill and low pay. NFU Scotland disputed the implication that those employing "low skilled" workers are using "cheap labour" (Written Evidence 2020). Employers highlighted the contrary, citing pay of £13/hr with seasonal workers not "cheap" labour' "because piecework earns many well above the National Living Wage" (EFRA Committee, p23). The government demonstrated little intention to amend its position on immigration or skill, again responding by extending the seasonal worker visa, whilst re-emphasising the longer-term goal of increasing domestic recruitment and automation. No strategy to recategorise horticultural workers as skilled succeeded; there was no shift away from prioritising high-skilled – meaning highly qualified – immigrants.

Government believed "the robots are coming" to replace the need for human migrants (Farmers will be told to replace low-paid immigrant workers with Robots, MailOnline, June 12, 2022). But experts were more skeptical about robot pickers:

> "Dextrous manipulation in unstructured environments is a big challenge for robotics today," said Dr Vishuu Mohan, from the university's School of Computer Science and Electronic Engineering. "Skilled humans find it effortless, but when we try to build a system which does the same thing it is a complex integration of vision, touch, force and movement and on top of it the ability to learn and adapt – which is the only way to deal with any changing, unstructured environment".
>
> (Fruit-picking robot designed to replace seasonal migrant workers on British farms will be 'ready by Brexit', The Independent, August 28, 2018)

Those within the industry became frustrated with politicians and others telling them they don't need seasonal migrants because British workers and automation are available right now (Addressing labour shortage 'not just a 2021 problem', The Western Mail, October 20, 2021). But they seemed to stomach the suggestion of this medium-term solution, in return for more seasonal worker visas now.

More seasonal workers – for now

When the EFRA committee again considered labour shortages in food and farming (EFRA 2017), the analysis emphasised near-term causes – pandemic and Brexit – with none of the rethinking of skills and food jobs as "low skilled" seen in the previous inquiry. Growers offered evidence of how the loss of EU workers had left vacancies despite higher wages, with better pay moving people between employers but not attracting new recruits. They expressed concerns with the new immigration system's bureaucratic inefficiency and expense, including the seasonal worker visa. Worker exploitation not much commented on since the early 2000s became more prominent as the visa was shown to create new risks for workers (FLEX Written Evidence 2022). These concerns arose partly from the government's inadequate review of the pilot which was not wholly independent and lacked data to assess risks against ILO indicators of forced labour. DEFRA's review did identify unacceptable welfare conditions and experiences of "racism, discrimination, or mistreatment by managers allegedly on grounds of workers' nationality by being subject to disrespectful language or passed over for better work or accommodation" (DEFRA 2021a, 2021b). However, no operators were removed from the scheme, and it was not substantively changed; the committee recommended the seasonal worker visa route be made permanent. Seasonal workers who participated in FLEX's research were the only ones heard across the enquiry, and they seemed to make little impact.

Having operated for a whole season post-Brexit and post-pandemic, growers felt able to evidence impacts of decreased European migration:

> In practice the industry has seen a significant reduction in the numbers of staff with settled or pre-settled status looking for work in the UK. In previous years these staff provide the back-bone of the seasonal workforce. Many returned to the same farm year after year providing important continuity and much needed experience.
>
> (British Growers Written Evidence 2022)

Emphasising the value of returnees and those working towards settled status subverted the long-standing argument that horticultural workers are not an immigration issue as they work a season and then go home. In fact, growers value returnees for their experience and productivity, and the majority of returners have settled status (Barbulescu, Carlos Vargas-Silva & Robertson Written Evidence 2022). Growers portray their employees as temporary presence, whilst really wanting them to be semi-permanent.

The committee and government continued to regard seasonal migration as a short-term measure until domestic labour and technology fulfil labour needs. Industry actors seemed to accept this vision as the quid pro quo for having access to seasonal visas, but remained skeptical:

> automation is unlikely to reduce total labour demand because whilst the industry is already making good progress in increasing labour productivity, any gains in productivity are being used to allow companies to grow their output to meet rising UK demand.
>
> (Greater Lincolnshire LEP Written Evidence 2022)

The government established a review on the topic, which suggested some technologies offering low to medium labour displacement are ready for mass adoption, with others expected beyond 2030, but "many tasks and crop groups may never be automated" (DEFRA 2022d).

Through 2022, Conservative administrations faced pressure to soften their line on immigration, with recurrent reports that DEFRA and Home Office disagreed on numbers of seasonal visas. The solution Liz Truss found was to pitch migration as an economic necessity to fuel growth (Truss's plan to relax migration splits the cabinet, The Times, September 26, 2022). In December 2022 more seasonal visas were finally announced for 2023–2024. It is notable that Truss avoided referring to these as low skilled jobs, instead stating the importance of "the right skills that the economy, including the rural economy", need to "stimulate growth" (Liz Truss plans MORE tax cuts in the New Year, MailOnline, September 24, 2022). It seems that even when the economy needs "low skilled migrants", they are unwelcome, whereas those with *essential* skills are palatable.

An expanded seasonal worker visa gave horticulture what it wanted, and lobbying eased. Then 2022 brought the "cost of living crisis", and continued pressure on food supply chains, in part from Russia's invasion of Ukraine. Challenges during 2021 led growers to warn of unpicked crops "left to rot"; additional pressures from higher priced inputs led them to threaten reduced production ('Cucumber capital' growers selling up as Brexit and energy crisis hits Britain's vegetable industry, Guardian, September 6, 2022). The invasion directly squeezed the labour supply chain as a significant proportion of seasonal visas have recently gone to Ukrainians. The plight of some workers stranded in the UK without family, work or visa drew sympathy (Ukrainian workers flee 'modern slavery' conditions on UK farms, The Guardian, April 19, 2022). Recruiters made up shortfalls from further afield in more economically marginal countries. Poor treatment of these workers was soon exposed, including recruitment fees which essentially engaged migrants in bonded labour. Recruiting in Southeast Asia, often at a speed, had seemingly circumvented checks on new overseas partners (Revealed: Indonesian workers on UK farm 'at risk of debt bondage', The Guardian, August 14, 2022). Sadly, those who had warned that a quickly implemented, weakly monitored scheme is open to exploitation, modern slavery or trafficking, were proved right.

Who picks for Britain today?

At the time of writing many, issues which made headlines at the close of the 20th century have returned, albeit in updated forms. As in other European countries, the pandemic only temporarily valorised "essential" horticultural workers, with little enduring impact on policy or treatment of workers (Carnibella and Wells 2022). Post-Brexit immigration rules prioritise workers with high skill levels, and shortage occupations, categories excluding horticultural workers. Seasonal horticultural workers are allowed special status as necessary "low skilled" migrants, controlled through dedicated visas and recruiters. The government remains reluctant to intervene to assist growers, seeing it as a matter for employers to create more attractive jobs or develop automation. The industry counters this by appealing to the national significance of its produce, and the distasteful prospect of higher imports, rising food prices or wastage. Attention focuses on numbers of vacancies, and quantities of migrants, not qualities of the work or the workers employers want. Workers who do the work may be less of a secret than they once were to consumers. But the veil over labour in the supply chain is rarely fully raised, rising not to showcase the skill of valuable workers, but to laud them as durable bodies, or expose them as victims of exploitation.

Across the past three decades, horticultural work and workers featured in public debate as archetypal characters. These draw on wider narratives of good migrant, bad local which have been around at least as long as Eastern European seasonal workers (Scott 2022). As revealed during the pandemic, employers draw on these narratives to articulate their needs (Scott and O'Reilly 2022), which centre on migrant workers as an economic necessity (Broomfield et al. 2022). Stakeholders recount stories about the labour force, a tale which always ends the same: we need more seasonal workers.

Most prominent in the media is the archetype of **beleaguered farm owner** or employer who just wants to get their Great British crops harvested, but is hampered by lack of workers and limited government sympathy. Victims of successive crises, they lose money, throw harvests away, suffer stress, having little power to improve their lot due to the power of supermarkets. Brexit is making their lives even harder. They believe the economy and society need them because domestic food production is simply better; Wimbledon needs British strawberries. What is a family Christmas dinner without British sprouts? But they are being pushed to the limit and it may no longer be worth growing in the UK.

Behind the farmer, receding into the background, is the **heroic migrant** who endures wind, rain and long hours to harvest crops because they just want to work hard. They appreciate the opportunity to earn well through speed, diligence and a favourable exchange rate. They are hardy and uncomplaining, whatever the weather, however long the hours, grateful of the chance to work in the Great British economy. Occasionally the hero who has made the UK home surfaces – their good work rewarded by promotion and settled status. But the most heroic migrants don't want to stay in the UK for good.

At times of crisis, migrants' heroism is celebrated, but at other times resented as an unwelcome foreign presence, a reminder of the **lazy Briton**. The domestic worker doesn't want to work in the fields as it is too hard, the pay too low and the accommodation not nice enough. Local recruits who give it a go are too slow and uncommitted. Current generations were once demeaned in contrast to their harder-working forebears. Nostalgia has been superseded by post-pandemic mythology founded in evidence of failed drives to recruit British pickers.

The inadequacy of the local workforce creates demand for the **shadowy labour agency**, specialists in overseas recruitment, acting distant from farms and the spotlight. Less the evil gangmasters they once were, they are now lax administrators. They provide employers with what they need, including acting as scapegoats when things go wrong. They have their own shadows – sub-contractors or sub-sub-contractors, including some who bend or break the rules, especially when far from the UK.

Also in the distance is the **ill-informed government minister**, far from the reality of fields and farm finance, with unrealistic expectations about immigration and labour. They blame the industry for its own recruitment challenges and seek to wean farmers off the addiction to cheap migrant labour. Ministers often spy a new character looming on the horizon, crying "**let them use robots**". If really pushed to intervene because supermarket shelves are empty, a temporary fix will do, but broader change must come from the industry. The Great British farmer should try harder to attract Great British workers by paying them better.

Both ministers and farmers claim to act in the interests of the **poor British consumer**, who just wants to pay the lowest prices for good British produce. In the past they were only figuratively "poor", meaning deprived of choice and good produce whenever farm labour was short. More recently they include those literally too poor to pay rising food prices.

These archetypes socially construct needs each actor wants to mobilise: for more labour supply, or less immigration. When events interrupt the ongoing discourse, they temporarily adjust arguments to reinforce the preferred construction: see we told you Brits couldn't do these jobs. The greatest indication of how entrenched these lines of argument are is how quickly people forgot that field workers do essential, difficult work. Seen from a longer view, pandemic-era discourse continued ongoing narratives, becoming just another in a series of labour crises. In the UK, as elsewhere, government responses focused on short-term solutions, not longer-term problems (Carnibella and Wells 2022; Corrado and Palumbo 2022; Milbourne and Coulson 2021). The clear silence is around the work's nature, what the workers do and how they accomplish it. Quantities of vacancies and numbers of visas gain far more attention than the qualities of jobs or of workers' experiences. Whilst mistreatment of workers occasionally features in public debate, the most prevalent moral concern is the prospect of crops rotting in the fields; labour shortages must be avoided to avoid immoral food waste. The immorality of treating workers as temporary presences with carefully constrained rights seems easier to overlook, particularly when workers' own voices rarely feature. The supposed will to protect Great British Produce serves as a nationalistic, biopolitical cover for demand for precarious labour from poorer countries.

Debate centred on archetypal characters reinforces a focus on individuals as the source of skills and labour issues, individualising problems and solutions. It is thanks to the heroic attributes of a migrant that they can do horticultural work, whilst it is the fault of deficient locals that they won't. Rather than seeing skill as practices developed in social contexts, it is assumed to be a capital individuals have or do not; responsibility for acquiring skills which gain someone the right to immigrate lies with the individual. Each business should make sure it creates jobs which people want, denying systemic features of agri-food supply chains which make this so hard, or any role for public intervention. In reality, the micro scale of the individual finds it hard to transform work conditions created by micro, meso and macro scale interactions, outwith their control (O'Reilly and Rye 2021). Fault and solution are positioned with individuals, to the neglect of systemic factors, or collective responsibility to shape economies in which good workers are fairly rewarded and treated.

Government blames unreasonable farmers, and vice versa, allowing both to overlook the need to tackle interlocking social and ecological crises which threaten food security and perpetuate injustices. By trading on archetypal characters and pitting them against each other, the discourse around horticultural labour skirts over deep systemic problems which underlie food systems reliant on mobile labour (Klassen et al. 2023). Minority reports were sidelined in parliamentary inquiries, easing pressure on the government to tackle entrenched food-system challenges. The "yet another crisis" narrative apparent across decades of public discourse reveals how actors focus on the latest problem as if a one-off, rather than seeing a series of inevitable failings, symptomatic of capitalism's inherent inequalities and food production reliant on flexible migrant labour.

The past 30 years of debate reveal that seasonal visas have never been a temporary fix, rather a symptom of enduring reliance on spatial fixes accessing labour in marginal economies. The UK economy is structurally dependent on a degree of low-wage immigration (Coulter 2018). Employers portray themselves as subject to a series of short-term crises in labour supply because it allows them to maintain access to their preferred labour source: temporary migrant workers who bring qualities of hard work and acquiescence (Campbell 2019). If horticultural jobs have to compete in the labour market, they lose out, hence preference for visas tying migrants to the sector. The industry is aware that it is more likely to be given access to migrants by highlighting them as temporary presence with limited citizenship, hence emphasising students working in the holidays. Given high reliance on EU citizens with settled status in the UK and growers' preference for those who stay for more than 6 months and return annually, temporariness is a carefully constructed myth.

Being able to control exactly how temporary jobs are is crucial to growers, a factor easily lost amongst debate over whether growers can pay workers more. Low pay is not necessarily what makes seasonal jobs unattractive, but can serve as a convenient distraction from other conditions. The broccoli pickers earning £30/hr gained notoriety, but were not an anomaly as experienced pickers have long been able to earn decent wages through piece rates or bonuses. But employers only want to offer work for as short a period as possible, hence temporary contracts, and deployment which gives the most productive workers more shifts:

> The industry requires workers who can commit to regular patterns of work during the season with the flexibility to work increased hours when crop or market conditions dictate.
>
> (British Growers Written Evidence 2020)

Migrant workers fit most neatly into flexible work regimes where employers retain flexibility to only pay those they need to work each day, and to decide shifts at short notice. Workers who borrowed money to travel therefore struggle to earn sufficient to cover their debts (Mellino et al. 2022). Flexibility for employers means precarity for the employees who lack control or rights and face bodily and emotional endurance (Scott and Rye 2021; Stead 2021). Migrant workers are more likely to accept precarious positions and, once employed, struggle to challenge conditions which easily slip towards exploitation (Gertel and Sippel 2014; Campbell 2019; Milbourne and Coulson 2021). Employers justify labour flexibility by portraying it as what workers want or blaming the market (Bonanno and Cavalcanti 2014).

Migrants' relative pliability was highlighted during 2020 when they worked alongside the domestic workforce. As noted in the previous chapter, growers told me their local recruits were more vocal about long hours, challenging practices such as curtailed breaks which migrant workers apparently accepted. One producer found enthusiasm for local recruits "was soon dashed" as most did not take jobs offered to them, declining positions because of not wanting to work weekends, wanting fixed shift patterns, only wanting part-time roles or "it was too hard a work" (Riviera Produce Written Evidence 2020). Most of these expectations seem quite reasonable and would represent decent work conditions from a flexible employer and would also be standard in many industries. But because horticulture has been able to operate without typical worker rights, such requests seemed exceptional and unreasonable. Producers seek to protect their access to flexible employment, meaning migrant workers in precarious roles.

What about skill?

When I set out to review media coverage of horticultural work, I expected signs of a shift towards greater recognition of it as skilled work, particularly as the pandemic foregrounded "low skilled" essential work. I have heard speakers at industry events arguing that anyone who knows horticultural work recognises it as highly skilled. In fields and polytunnels, growers echoed this, pointing to fast effective workers: "I could never do that". Other industries made exactly these calls during their 2021 labour crises, seeking to be counted as high-skill shortage occupations. But horticulture has rarely highlighted itself as a skilled sector. Even when the introduction of skills-based immigration policy incentivised this, not all voices highlighted their employees' skill, with some preferring to emphasise that even low-skilled roles are economically essential or note the value of "experience" rather than skill per se. Perhaps, like the employers of the early 20th century (Verdon 2017), growers were wary of fuelling wage inflation by acknowledging their workers' skill. Once growers secured more seasonal visas, skill largely disappeared from their narrative, demonstrating how they construct workers and work to serve their preferred solution.

In media and policy discourse, horticultural work remains associated with low or unskilled work and the same categories of migrants. When Lord Lawson made his much criticised comment that we don't need Romanian fruit pickers but highly skilled City people, it was not his distinction between high and low skilled work which was challenged. The focus was numbers – of people, workers and migrants – not their abilities, which remained hidden by the moniker "low skilled migrants". In press coverage which likes simple caricatures such as the heroic migrant worker, there is little appreciation of the complexity of tasks or workers' skill; it is sufficient to portray them as knuckling down to hard physical work in tough conditions no local will endure. The one parliamentary inquiry which did question whether horticultural jobs are accurately categorised as low skilled did not conclude that they should be re-categorised, or that the government should rethink the grounds for designating roles as skilled. Some challenged the fallacy that migrants on farms are unnecessary, but not that horticultural workers are unskilled. This suggests defining skill in association with formal education and high pay is deeply entrenched in how we think about work and learning, and in social categories which rank occupations.

Horticultural employers are well aware they benefit from seasonal workers' skill, but this is more likely to be described as the value of "good", "experienced" or "specialised" workers, able to work carefully at speed. A phrase the NFU used recurrently between 2016 and 2021 to describe the ideal horticultural workers is "reliable and competent" (Brexit vote brings uncertainty for fruit pickers and farmers, The Guardian, October 16, 2016; NFU 2017; UK could run out of food a year from now with no-deal Brexit, NFU warns, The Guardian, Aug 7, 2018; Addressing labour shortage 'not just a 2021 problem', The Western Mail, October 20, 2021). It seems farmers want reliable, competent workers, not skilled ones. Perhaps they know the skill of a good picker can be acquired on the job, associating skill with the job rather than the person, so people are interchangeable. Or perhaps the ability to fit into flexible workplace regimes is the skill farmers really value.

The exceptional context in which growers portray the skill of seasonal workers is to rebut the suggestion that robots can replace them. In the NFU report calling for access to "reliable competent workers", horticultural workers are only described as skilled to challenge policy makers who regard automation as their imminent substitute:

> many crops such as berries, apples and pears require skilled hand-picking to avoid damaging the fruit. Technology has not yet been developed to replace human pickers at an economically viable scale.
>
> (NFU 2017)

The scientists developing robots to emulate these abilities know that fruit picking entails dextrous tasks which robots struggle with, whilst "skilled humans find it effortless". Similarly, when growers sought to attract UK recruits, it made sense to portray the jobs as complex to make them more appealing:

> "It's not rocket science but it is quite a skilled job that these guys do," he [farm owner] says. A top picker needs to know and apply supermarkets' exacting standards for exactly how fruit and veg must look, as well as how to cut it and pack it.
>
> ('Feed the Nation': Britain urged to take up work in the fields as coronavirus threatens fruit and veg supplies, The Independent, March 27, 2020)

Such anomalies suggest that employers portray horticultural work and workers as skilled when it serves their needs; most often they need flexible, productive labour.

Conclusion

The recent history of public discourse around horticultural work highlights that what is portrayed as one short-term crisis followed by another, then another, is actually systemic reliance on temporary migrant workers. Horticultural employers seek their ideal workers, sketching the comparative qualities of domestic and international recruits to secure government support supplying their preferred workforce. They struggle to recruit within the UK because of the jobs and conditions they offer – the quality of the work, but stakeholders trade numbers of unfilled jobs, or how many migrants can be allowed in, focusing on quantity of horticultural workers to hide matters of quality. Growers are challenged to pay people more to attract the workers they need, again assuming a problem of quantity – of money – when it is flexibility and precarity of jobs which are the qualities employers really need. Public discourse too easily gets distracted by questions of how many and how much, overlooking the questions really needing attention: why are food growers experiencing crisis after crisis, and how can we stop the burden of this falling on workers?

This chapter has shown how horticultural work, as supposedly low or unskilled employment, is treated in UK immigration policy, demonstrating how skills-based approaches construct skill and shape hierarchies of work. Skill is interpreted and applied to designate social categories which deny continuity between "intellectual" and "manual" work, demeaning "low skilled" jobs and workers. These tendencies contribute to the devaluing of horticultural work and have been perpetuated by actors wanting to retain access to flexible temporary workers.

The focus on quantity rather than quality of work, and skill-based immigration rules, both individualise work and workers by focusing on traits of archetypal characters and jobs, rather than considering connected systemic pressures which shape employment practices. Skill is regarded as a capital which can exist independent of human practice and learning, as something which bodies either have or don't, neglecting its situated and relational features. Skills shortages are regarded as deficits of abstract expertise, rather than the result of flawed systems which impede flows, and block people from becoming growers. The myths and misunderstandings about skill which circulate are too easily believed because of lack of attention to what horticultural work and skill are really like. So what exactly are the skills of horticultural work?

Notes

1 The database LexisNexis was searched to identify stories including the terms horticultur* work*, seasonal farm work*/labour, SAWS/seasonal agricultural workers scheme. Results were filtered to include items published in the UK and referring to geographic context of Europe.

2 Labour shortages in the food and farming sector 2021–2022; Covid-19 and food supply 2020; The UK's new immigration policy and the food supply chain 2019–2021; Feeding the nation: labour constraints 2017 and 2018. For each inquiry, qualitative analysis was applied to committee report(s), government response(s) and submitted written evidence.

3 All written evidence submitted to this enquiry is available at https://committees.parliament.uk/work/2539/feeding-the-nation-labour-constraints-inquiry/publications/written-evidence/.

4 All written evidence submitted to this enquiry is available at https://committees.parliament.uk/work/1497/labour-shortages-in-the-food-and-farming-sector/publications/written-evidence/.

5 Knowing Good Growing

What does it take to grow a plant? Connecting inward and outward worlds

Google or a childhood science lesson suggests all plants need seven things to grow: space, the right temperature, light, water, air, nutrients and time. For horticultural enterprises, this simple process is incredibly complex, but growing plants is only a small part of what they need to know; what's key is how plant growth interacts with other processes. Across the growers I studied, I wanted to identify fundamental skills across type and scale, unique to businesses founded on plant crops. A favourite question was: on a scale of 0–10, how much did you know about growing plants when you started your career and how has your position on that scale changed? Explaining movement around the scale prompts reflection on learning, often revealing how people consider horticultural skills. This manager at a large organic grower told me what he knew when starting out and what he had learnt since:

> So I knew what a seed and a plant was. They go in the ground and they need water, and you need some other folk and a tractor and a van to sort it out. But you know, expanding inwardly into the granularity of the detail – inwardly and outwardly I would say. Oh, and understanding the universe you're operating in
>
> (L1.2)

He encapsulated what commercial growers need to know and how their expertise sits at the intersection of the plant with everything around it, which must align to get products to customers. Looking *inward* to the plant growers need to understand the "granularity of the detail" of what is happening with growth and how to respond. Looking *outward* they consider the wider operation and "the universe" that influences it, not least what customers want. A grower's knowledge stretches from close-up plant work, across supply chains, at meeting points of macro, meso and micro contexts captured by an ecological interpretation of skill. Here I start with inward looking, plant focused elements and work outwards to external conditions. Getting it right is both incredibly simple – plants

DOI: 10.4324/9781003281689-5

just grow – and highly complex, with growers balancing their ability to coordinate plant growth, with their inability to be wholly in control.

Tending plants by attending to conditions

Plants' seasonality shapes the working year. Take soft fruit growing:

> So there's sort of a typical year when we start in January. We're starting to plant nearly as soon as we come back and planting carries on through to April. Then once we get to sort of the middle to the end of Feb, tunnel plastic starts to go on, fleece starts to go on. Once planting's in, we then cover everything with plastic, we'll fleece certain bits, not everything, again to sort of help us with our profile, our picking profile.
>
> When we get to this time of year fleece is coming off, there's a lot of husbandry work to do. Then it starts to warm up, so irrigation then becomes key. Obviously, they don't need a lot of water earlier on, but they will need the food, so it's getting the feed into them. And then as soon as we start seeing the flower, fleece needs to come off and bumble bees need to go in to start the pollination process. It's – that's quite a crucial time when they're in flower, the weather plays a big part. So obviously if it's too hot you don't want it to be too humid in there, so you need to vent the tunnels. Where if we go into a cold period and we've got frost coming, it's putting the fleece back over to protect the flowers again.
>
> And then the sprays start, generally with a spray window of every seven to 10 days, once the fleece is off, depending on what we're seeing in there. So we'll usually do one round of fungicides and then the following week will usually be around the sort of foliar feeds and bio stimulants. And that sort of carries on through to picking, depending on [what] the plant's and whether it's a June-bearer or ever-bearer. If it's a June-bearer we do a lot of truss work too pulling the flowers out, to present the fruit. There's a lot of it, and it helps with disease control and picking speed. So it's try and present as much as we can and then ever-bearer it's more – we don't do truss work there, but it's more managing the plant so it doesn't get too big. So you might do some de-leafing during the summer to sort of keep that plant at the optimum size. So again, that's sort of helping with spray coverage and picking costs and they haven't gotta rummage through a load of leaf to find that fruit at the bottom.
>
> So it's sort of – that's an ongoing process and then obviously the irrigation ramps up through the year as well and becomes very crucial in that summer period. So we have irrigation probes in all the fields, which relay information back on the Cloud every half an hour. And then once a week we have a meeting with an irrigation company who look after all these probes for us and talk about what we did last week, what we need to do this week with the weather forecast we've got and we're looking at the EC, the pH of the water that's running off, that's going in, and the EC of all

> of that to manage the bag, the growing environment for that plant at the stage it's in. We have two [irrigation] tanks. So an (A) tank is your calcium and your iron, and then your (B) tank is your potassiums, phosphates and all your micro elements, so your boron's, your copper, zinc, all of those bits. So we send off once a month samples of the coir, the leaf, the run-in water and the run-off water, that's analysed and the feed recipes adjusted once a month based on those values then and what the plant is using or not using or running short of. That's done by our agronomy company.
>
> And then, when we get to the end of the year or the end of the crop, depending on whether we're going to over-winter or not – but we don't over-winter a lot anymore, from a disease and pesticide point of view. So the plants then will get pulled out, plastic goes down, if the bags are three years old we chuck the bags away. That sort of brings us sort of towards the end of the year then. And then it really does quieten down in December. (L3.2)

Growing starts with a seed, ensures the seven essential inputs – eight if the plant is not self-pollinating – and prevents pests or diseases from doing harm. The challenge is optimising all inputs and regulating conditions to produce consumable fruit, leaves or tubers on schedule.

Commercial producers do basically the same as home gardeners:

> So like your general back garden grower will probably be able to have a nice little flush and then it'll probably have a lot of little, small ones but they'll still be red and they'll probably still taste the same. But it's having that appearance and the quality of good quality fruit throughout the season, which is achieved by managing your plants and doing all these little husbandry tasks and looking after your pests and diseases to give you that visual quality that that fruit is good all the time and not got any defects on it and that sort of thing, and tastes, tastes good throughout the season. (L3.2)

By understanding how plants respond to conditions, the grower manipulates them into growing to the preferred rate or form. Growers assess environmental conditions, understand how a plant will respond and act to modify them (Dixon 2016). The skill is attending to the environment, assessing what is happening and then responding in a way attuned to that moment's needs (Ingold 2000).

Expert growers can adjust inputs more precisely, generating more predictable outcomes. This begins with selecting optimum plant varieties to achieve a crop with the best flavour also to aid production schedules:

> Because of 12 months' supply, varieties come ready at different times. So we probably have about 30 to 40 different varieties of cauliflower, they have different maturity dates and we plant them at different times. Also on different geography and different aspects and different climatic areas

> to ensure they're all come ready at different times. But even in the nicest way possible, if we have a large spell of drought, all the crop can come – a lot of crop can clump up together and then it's a write-off.
>
> (L4.3)

Larger operations develop close relationships with plant and seed suppliers, so they can access new varieties with "the right genetics" as one plant raiser put it. Customers are an influence but do not have the final say:

> So it's a toss-up for us for us to say, 'well yes, that tastes and looks nice, but it yields crap and it's horrendous to pick'! Before we put it in front of them [supermarkets] we've gotta know that we can grow it and it's easy to grow and it can pick well and then we'll put it in front of them.
>
> (L3.3)

The breeder has the know-how to select varieties to meet their needs (Graddy-Lovelace 2020), but they are not the only skilled actor: the plants are sensing and adapting to conditions, responding so they can flourish (Ingold 2011).

The next crucial input is water with irrigation rates determined by plants' varying needs:

> So that's actually nice and damp at the moment. Cucumbers are quite shallow rooted, so they need regular watering. Because their roots just don't penetrate very far down. So if you give the soil a really good soaking, and it soaks really far down, it's kind of wasted water. So we're better off doing little and often. But tomatoes, you can get away with more so, it might be that the cucumbers you give a bit of water every day, and the tomatoes you give extra water but every other day. Which again is a little bit more efficient. But again, you also have to keep it even as possible. Because again, tomatoes, if you gave them hardly any water for a week, and then gave them a load to make up for it, actually there's a chance that their skins will burst and the skins will split because they've had too much water, and suddenly, the tomatoes, you know, push out too much, grow too quickly.
>
> (L1.3)

They can't water all plants to a routine schedule because conditions and plants vary; the skilled grower attends, judges and responds:

> if you get your irrigation wrong the plant will go into stress, then mildew comes in and disease comes in and then fruit quality drops. So irrigation is probably the biggest, the most important thing.
>
> (L3.2)

These strawberry plants are in an inert coir-based substrate through which feed solution is flushed to optimise nutrients and moisture; the skilled grower enrolls materials to assemble conditions favourable for growing (Ingold 2000; Krzywoszynska 2016). Outdoors, weather is still the grower's greatest enemy; they cannot control it but seek to understand its flows and bend with them (Ingold 2011). Working with uncertainty is the skill of growing (Krzywoszynska 2016).

Water delivers strawberry plants feed promoting growth and later, nutrients stimulating fruit production. The same principles apply to field crops, but interactions between plants, soil and inputs are more complex:

> changing the amount of nitrogen applied, just as one example, changes not just the size of the crop you grow, but also the health of the crop. [Understanding] how that relates to shelf life, how it, you know, in some instances relates to taste, and then growing or working with the growers to understand what balance they're looking for. So, you may produce a smaller plant, but your costs may be reduced and you may end up with, you know, greater soil health through not applying these fertilisers, which will benefit the next crop and so it's all about understanding how those different variables interact with one another.
>
> (L2.5)

In this large company, the scientifically trained trials team test these interactions and make cost-benefit assessments to identify ideal input rates. Other growers combine experiential knowledge with that of experts and technology (Brice 2014; Oreszczyn et al. 2010), sending soil or plants for lab testing, but relying as much on judgement and practices such as rotation to maintain plant nutrition.

Organic growers are particularly conscious of soil health:

> I think, as an organic grower, understanding soil and the strains you're putting on the soil, if you like, and also how you can alleviate that and make better soil, I think that's really important so being able to make compost, being able to understand why you want to make compost, being able to understand why plants grow or they don't grow, having some basic understanding.
>
> (S1.1)

This grower converted a conventional livestock farm to organic horticulture, but farming neighbours were skeptical, judging the land too poor for growing:

> using the older farmer phrase, I was coming off the 'boy's dirt' in Hampshire where a yard of topsoil, easy weather, bright sunshine, no Atlantic gales, whereas you turn here and the sheep look at the garden and go 'really'? It's like this is 'man's dirt'. It's much harder, much more difficult.
>
> (S1.1)

He sent the soil for analysis, got advice on adjusting the pH and then did the "man's work" of slowly improving soil fertility; he attended to conditions on his farm, then shaped them to favour growth.

Space, air and light – Assembling infrastructure

Space, air and light are interconnected, shaped by materials assembled to regulate them and provide plants room and support to grow. Support such as twine hanging through polytunnels, which plants are looped around:

> With this, you're training them up. The cucumbers, actually, you're doing this more frequently, partly because they grow even faster than the tomatoes, but also partly because the tomato plants are slightly sort of firmer, woodier stem. So that one there, you saw, it was drooping right over. And they'll do that a lot faster. And, and if it's sort of drooping down here, then it becomes really slow to do your side shooting and twining, so then you'd have to do that and take a lot more care so when you're doing your twining here. It does take a little bit of time to know how far to push it, because you want to twine it as much as you can, because the better you twine it, the further you can leave it till next week. But if you try to twine it round too tight, you can actually snap the top off and then that's it, we've lost this one.
>
> (L1.3)

This instruction in twining cucumbers highlights plants' fragility, and how many tasks are intended to make harvesting more efficient by ensuring produce is more visible to pick.

Spacing plants allows air circulation, but too much is not desirable, hence polytunnels and other cover "that stops the crop from being hammered by wind, and heavy rain" and "keeps the environment much more stable" (L1.1). Covers allow growers to influence conditions: "with tunnels we can try and manage it [weather] as best we can" (L3.2). High-grade polytunnel plastic lets UV through; white truss tape and plastic ground cover also maximise light reflection onto the plants. Through specialist materials, growers acquire knowledge that went into designing them; situated in new spaces, the materials carry knowledge which becomes part of the context where growers learn (Sørensen 2009).

Light and air circulation are also maximised through plant husbandry. For strawberries, this entails de-runnering and then de-leafing to manage the plants' leaf area index. How do pickers know what to remove?

> Every runner needs to come off, so we do that on a piece-rate [work] and take them off. De-leafing: in all tasks we do you've gotta try and make it sort of idiot proof and sort of as simple as possible. So we've – there's a bit of truss tape that goes down the side and most of the de-leafing you

> say, anything that's touching, or on the truss tape, take that off. And that gives us that sort of evenish balance of not too much on some plants and maybe just slightly not enough on others, but overall it's about the right balance. So it's pretty hard to sort of say, 'oh well this one needs three leaves off but actually that one only needs one'. And from a cost management point of view as well, if we did it like that and treated every plant individually, our costs would spiral. So we've gotta be a bit realistic and go, 'right, anything on a truss tape, take off', and put that on piece work and off they go as well [...] So you can go and take too many leaves off and that can have a negative impact on your crop, you'll suddenly have smaller fruit, so your yield will be lower. So that's pretty crucial.
>
> (L3.2)

Pickers are not expected to know what a leaf index is and lack time to calculate it for every plant. They need to know what to attend to – leaves touching truss tape – and how to respond – remove leaves – so the expert crop manager developed a "rule of thumb" to communicate essential know-how derived from experience (Krzywoszynska 2016). In workplaces with a strong division of labour and rule-based instruction, pickers have limited agency to judge what plants need and little opportunity to acquire plant know-how.

Pickers are trusted with crucial plant work, meaning they have power to harm tomatoes:

> we'll pinch out the side shoots because they are concentrated to grow upwards and you end up with a thinner plant so it's better for ventilation, better for air circulation, but if you pinch out the growing tip, the top of the tomato plant, it will just stop growing and so it will only get to two feet tall instead of eight feet tall and you'll only ever get 15 tomatoes off it and if you do that wrong at the beginning of a season that's it and you haven't killed the plant but it's not going to grow right.
>
> (L1.3)

Perhaps workers given less agency to practise and develop know-how are more likely to make mistakes, particularly if time is too tight for managers to communicate the importance of getting it right. Workforce organisation and ownership structures influence skill development as growers working on plants they have paid for, and from which they receive the profit may be less likely to kill them.

Workforce structure also influences how plant infrastructure is designed. Plant husbandry makes fruit more visible and, therefore, faster to pick because speed is paramount. Sometimes efficient picking does not align with optimal plant conditions, like the move to table-top strawberry cultivation which means plants don't benefit from ground warmth:

> if these were on the ground we'd be down on our knees and they would be shuffling along on their knees. 'Cos these would be on mounds that were on the ground and it would be back-breaking, absolutely back-breaking.
>
> (L3.1)

Those who pick this way in California call strawberries the devil's fruit because the work is so punishing (Papadopoulos and Fratsea 2018). Most UK growers set the cost of establishing table-top cultivation against reduced picking costs and opted in favour (Calleja et al. 2012). An effective horticultural business weighs multiple variables to judge optimal growing conditions, looking inward to the plant and outward to supply chains, particularly labour costs.

Friends and foes: Pests, diseases and weeds

Growers need expertise in foes which prevent plants from accessing essential inputs by shading or diverting water and nutrients. Planting seeks to reduce these risks:

> So the rotation's primarily trying to stop disease and carry-over of disease. If we grew cauli in here next year you'd be pretty certain we'd have club root which is where the root ball, rather than growing into loads of little roots it grows into four or five big bulbous things and they just can't take up the nutrients because they haven't got the surface area. The growth would be stunted, you'll find they can't cope with any stress, so hot weather they can't cope with and they just flop down.
>
> (L4.2)

Regulation has reduced the use of chemical controls which supposedly outsource expertise from the companies supplying them (Morgan and Murdoch 2000). But even farmers reliant on such inputs have specialist knowledge and adapt technical information to their land (Ingram 2008). Those growing undercover are particularly alert:

> if you have a problem with pests or disease it can spread like wildfire so that's a real one of being able to spot problems early on and being able to make the judgement call of if you've got a little bit of a pest damage, what is that pest? Is it actually merely cosmetic or could it actually affect the crop, then deciding what to do on that.
>
> (L1.3)

Assessing the plants' prognosis is a key skill, centred on attending to conditions and judging the best response.

A smaller-scale organic grower explained his pest control system almost sorts itself as he's created hedges which act as beetle banks:

> So if we do suddenly see there's a huge influx of aphids, which does occasionally happen, I can literally just walk into there and pull out a hundred ladybirds. But if you leave this here, which we've done for years now, what we find is that we've just got great cover for hoverflies to breed in. We just don't end up with a problem, which is great.
>
> (S1.1)

Learning how to attend to signs of attack is a skill supplemented by expert advice such as integrated pest management (IPM). This approach is popular with organic and conventional growers as it addresses resistance to chemicals and uses less environmentally disruptive controls (AHDB n.d.). Cucumber and tomato plants at the height of productivity demonstrated IPM in action:

> The idea is you'll have this background level of beneficial insects, so that when any bad ones come in, they're ready to gobble them all up. So we got some beneficial activity, so that one there, it's what we call a mummy, and that's a dead aphid. And what's happened is you get parasitic wasps, which are one of the beneficial insects we put out, and they come along and they will lay their egg in an aphid. And it will hatch out and eats it way out, and kills the aphid, you know, eat it alive from the inside. We also have predatory wasps, which do a similar sort of thing.
>
> (L1.3)

The wasps had been sent by the IPM company in vials and sachets hanging dotted around the beds. Tiny insects arrive in shakers which workers distribute around the plants, particularly on any showing signs of damage, like webs of red spider mites. The vials move insects around the country, carrying expertise which becomes part of the space where growers apply horticultural know-how. Near the polytunnels, I noticed a laminated poster displaying photos of various insects, divided into halves labelled "Friends" and "Foes", indicating to pickers which to be concerned about, and which are "one of the good guys!" (L1.3). Non-specialists provide essential "eyes on the crop", attending to conditions and then passing information to managers who respond with appropriate controls.

Organic growers work hard to remove weeds – plants "out of place" amongst crops (Mabey 2011). Mechanised weeders are used where crop spacing can accommodate a tractor, but hand weeding remains significant. During an autumn visit, I chatted with a team on their second day weeding an organic leek field; although they joked and smiled, the mental and physical strain was apparent, particularly when the supervisor mentioned they had at least one more day to go. The weeds weren't caught early and so became established:

> That's where that expression comes from, 'Hoe them before you see them,' because you're trying to kill those seeds at the beansprout, not even bean sprouting, it's smaller than that.
>
> (L1.2)

Growing then clearing the weed burden depends on the right timing, a feature of crop work which must coordinates with plant time (Brice 2014). The weeds should have germinated but not become established: "if you miss that window of opportunity you miss the easy way of dealing with your problem of weeds" (L1.2). But timing is tricky because growth's multiple variables are unpredictable:

> My manager called it 'cropageddon'. And what happened is because you can't prepare ground, you can't plant on time, the crops that are being planted, you can't weed them, so they get all weedy, they get high levels of disease. All your nutrients washed away, and we can't just go over with a fertiliser spreader and put Nitram back on again. You know, we have to rely on sort of soil generated nitrogen, and it really is a bit of a disaster really. It's difficult for conventional growers as well, but it is particularly difficult for organic growers, because quite a lot of our weed control methods rely on cultivation and so on, which you can't do in the wet weather.
>
> (L1.1)

To avoid cropageddon, skilled growers assess, then guide variables towards favourable outcomes; this requires knowledge of the eight things a plant needs, how they interact with the plant and with each other. The grower "looks inward" to the plant to notice these interactions, using skills of attending and analysing what they have noticed to judge how to respond. All this must be coordinated with "looking outward" towards the end product, what consumers want, how much they will pay and how much of that price the grower can spend on production. They constantly connect the inner world of the plant to the outer universe they operate within, all complicated by the unpredictability of plants.

Attending to detail

A skilled grower may not know how biochemical processes work in the way a botanist does, rather they apply experience with similar plants. For example, judging how many leaves to remove to allow air and light onto tomatoes:

> You could take six or seven off, it won't kill the plant but it will stop growing for two weeks because it will be effectively shocked and traumatised and a lot of that is kind of done by experience as much as anything. It's done by doing it the wrong way and then not doing it that way again.
>
> (L1.3)

The complexity of variables means being "a good grower" entails constant vigilance of plants' varied behaviour:

> It's going out and looking at your crops on a daily basis and sort of taking a feel of how they're growing and how they look and sort of going 'well actually something's not right here' and getting down to the detail of what the problem is and sorting it out as quick as you can. Like now we're on table-tops, irrigation is sort of massively critical. When I first started we were growing in the soil and if you didn't irrigate for three or four days it didn't really matter. Whereas now, if you don't irrigate for two hours on a hot summer's day you're already putting your plants under stress. So it's sort of looking for all those signs and making sure everything is – and all your people are doing what they say they're doing. So it's a lot of just monitoring what's going on out there and what you think's being done is actually being done to sort of keep the crop as healthy as you possibly can to give you the best fruit for that whole season.
>
> (L3.2)

This manager keeps an eye on the plants and for signs that workers have done what is required of them. Responsibility for plant health is distributed across teams, making coordination, communication and supervision crucial skills for deploying plant know-how across the workforce.

A good grower constantly monitors, noticing things about plants – a dusting of mildew on a strawberry, spots of leaf damage on a cucumber, traces of aphid husks on a tomato. Walking the crop is a common way for growers to monitor and notice problems, either themselves or via specialist consultants (CALU 2011). One grower referred to having "a radar" as they walked along lines of plants, attuned to traces of pests. This skill is acquired through education of attention, learning to notice what is important about others (Ingold 2000). With practice, growers "get their eye in" so they can spot what plants are doing, distinguish weeds from crops, or differentiate natural yellowing of leaf senescence from that caused by pest or disease. The old farming notion of "eyes on the crop" is quite literal, scanning for signs of plant stress:

> You'll have fruit ripening early, small fruit, leaf colour won't be sort of nice and bright and vibrant, or the leaves will be wilting or starting to get yellow or you go in, you have a good crop and a bad crop next to each other, it just looks paler and not … just visibly it's like, 'oh, something's not quite right there'. Yeah, they, they just don't look a happy plant, the leaves might be wilting a bit and, yeah, you haven't got nice bright, big, bold flowers, it's all the flowers all getting a bit small. So you can tell that something's wrong somewhere.
>
> (L3.2)

Growers educate their attention through practice, also through attending to and then imitating a skilled teacher, synchronising body and intentions with the expert in an empathetic engagement with people and place (Gieser 2008).

Posters like the one illustrating "friends and foes" help, as do visits from the agronomist to walk crops together. How does the expert entomologist guide a grower's attention to aphids:

> Well, they will look for aphids' skins, which is a sign that the aphids are there and they've been molting and multiplying. They would look for plant damage, of course, like the twisting or just deformation of leaves and growing tips, and the presence of sooty mould as well, which grows on the honeydew produced by aphids. And they'd also want to be looking out for presence of parasitoids and predators as well to see whether the native bio controls have moved in and how effective they are.
>
> (Sh4)

Crop walking is also fundamental in indoor growing:

> What you're doing in this crop walk is you're checking crop health, you're checking for any deficiencies, any unregular growth patterns, you're checking the developmental stages of plants. And whilst you're doing all this, you're also doing the same for the root culture and making sure the root is healthy and making sure it's, like, white fluffy as it should be. Cause obviously, the first indication to plant health is normally with roots if it's a nutrient slash pH issue. Obviously, if it's environmental, then it will be your leaves which'll be showing you the effects.
>
> (CEA1.3)

Having spotted pale-yellow leaves indicating iron deficiency, nutrient input is adjusted to make the plant happy again – a degree of control not possible outdoors.

One crop manager told me they go through each field two or three times weekly to estimate yields and look for trouble:

- When you're looking for mildew, what does that look like?
- Well you get lots of little white spores, like you'll have an area, normally on the back of the leaf, that might be quite small, like half the size of your little fingernail, that will have about 200 little white specks on it. And that's the beginning of having mildew.
- Can you see that from a distance or do you have to get right in-
- Yeah, because when it gets bad, on the front of the leaf, it will go yellow. And the back of the leaf there will be lots of little white spores. And when the weather is sort of mildew-y, it spreads quite quickly.

- Does that mean damper or -
- Yeah. Overall damper and maybe more humid. Yeah, I'm not exactly sure on the specifics of it

(L1.7)

What struck me about this explanation was that despite working at this farm for more than 20 years, plus having a degree in agriculture, he was not sure how mildew develops. Other times when I asked about causes of a blemish or malformed leaf, he could not explain; it was sufficient to know what was important to watch for and how it would affect the yield. This knowledge was not founded in biological insight to fungus, but through observing crop development, comparing plants and seasons. He suggested the key is being "quite focused":

> Focused on the cropping that you're producing. And getting a bit anal. On the details. Because some people are not. A lot of farmers are a bit like 'oh that'll do, just bung it in there, and just…'. But on an organic sort of intensive enterprise like this, you gotta – you gotta get into the details.
>
> (L1.7)

Details include preparing weed-free ridges in the beds, sowing the seed to achieve crop density crowding out weeds and checking the seed weight so the drill can be suitably adjusted, all of which depend on the drill and tractor working. So many details to get into.

Skilled growers have educated their attention to notice what is going on with plants, then assess how to respond. For those with capital, technology can take over observation and data processing, through monitoring equipment. Automated dosing in irrigation systems or temperature-regulated ventilation are relatively simple technologies. More sophisticated systems use extensive data and predictive technology, as this technician for a large grower explained:

> It's a sort of model basically, looking at heat, light, and historical data to predict, basically when a crop's gonna be available. And then we can kind of manage demand – potentially manage the crop in accordance with that so we can sort of try and match up our supply curve with our demand basically. So that's really important. There's Agrii, which is a Hummingbird: we use them to fly crop and map it basically. And you look at sizing, look at variability, you kind of assess crop that way.
>
> (L2.3)

The scale and specialisation of intensive production made it worth investing in these systems, working with companies like the brands named to tailor them. Prediction is otherwise by eye and judgement, informed by experience.

Attending to plants and deciding how to respond are the fundamental skills of growing, with desired outcomes achieved through care and judgement adjusting actions to suit the situation (Pye 1968). Growers' accounts of their expertise emphasised the many factors to monitor and analyse, for example:

- To have a type of mind that can factor in all the influences and come to a conclusion of what's best to do in … in the circumstances, which is just a sea of variables and then you have to cut the right path through them.
- What do you mean by the … the variables?
- Weather, people, machinery, orders, to name four

(L1.2)

Plant studies often highlight how plants are never wholly controllable or knowable (Lawrence 2022), but those who produce plant crops at scale strive to control and predict their activity (Argüelles and March 2022). Even complex growth variables have patterns:

> not every day is the same, don't get me wrong, but you can learn to predict what's going to happen, for sure.
>
> (L1.7)

But because plants are always up to stuff (Myers 2015), control is never complete, meaning:

> I also think problem solving is a really key part of the job. You – especially with crop identification or, like, even when you come to a fault and you might not understand it at first, being able to keep a cool head because it can be easy to panic quite easily. But also just being able to take in the information and learn from your mistakes I think is a really big part of it.
>
> (CEA1.3)

It is not good enough to surrender to the whims of growth; growers must assess likely trajectories and consider which actions will most likely "cut the right path through".

The complexity is clear from this analyst at a large grower asked to explain why a particular salad plant is their most challenging crop:

> Well, from a growing perspective, it's really hard to get right. As I said, uniformity is incredibly variable. You know, you can walk into a field, and there'll be heads of all different shapes and sizes, which makes harvest a challenge, you know, in getting it right in terms of timing, meeting customer requirements. It's also very volatile in terms of growth, in terms of weather events. So it can be very small, it can grow either very small and tight or, in a hot – as we've kind of had, recently – hot weather followed by a thunderstorm, it can really balloon in size, almost overnight. So getting

> that right is really challenging. I guess agronomically, it's tough. It's just that a range of pest, diseases. Herbicide bans have made weed control a challenge. And yeah, I guess from a commercial perspective, it's equally challenging 'cos it's very much viewed as an entry level product. Which, you know, means it's challenging on prices. There's a big price-match on with a lot of major supermarkets, which are really driving price down and made inflation hard to come by. Yeah, it's tough. And then, you know, in the meanwhile, it's because of that position as sort of the breadbasket or the base of the category, it means availability and reliability are – there's a lot of pressure on that from customers, whereas, you know, other slightly more niche products you might get away with a bit more [...] It's a real eye opener isn't it, when you get to Tesco, you see a lettuce or whatever on the shelf for 43p, there's – well, like with any product – there's the amount of research and challenges involved is infinite.
>
> (L2.3)

They need a predictable crop which should be homogeneous and calculable because this is industrialised agriculture (Argüelles and March 2022). The raft of growth variables is overlain with crop economics to ensure costs do not outstrip price, detail growers must be skilled in understanding and directing. Larger operations distribute responsibility for these variables across roles, but some growers cover everything – inward and outward views from plant to world.

When growers spoke of crops being challenging, they typically meant economic dimensions. The salad crop is challenging because the cost-price balance is so fine, hence considerable data processing. Another grower singled out cauliflowers: "the only way to make money growing cauliflowers is to not grow cauliflowers". He explained:

> So to get a crop today that's ready to cut, we'd have to allow 5% or 10% of the cauliflowers to be too big, to be even worth coming in. 'Cos it's not like many other types of crops, they're not uniform. They come relatively evenly compared to what they've done historically, but even still, they're not uniform. Any cauliflower that's too small or too large, may be perfectly sound and great to eat, hasn't got a home.
>
> (L4.3)

Financial precarity provides impetus to improve practices, as noted by a grower who supplies cauliflower plants:

> Well it's just a case of your pocket will tell you. The best thing in farming and being self-employed, you learn. The quickest way you learn is when you spend your own money and you make your own mistakes and then you don't make that mistake twice. And when you work for somebody

> else and you play with somebody else's money then, you know, it doesn't hit you so hard. But when you feel financial hardship because you've messed up, you make sure you're not gonna do it again, aren't you?
>
> (Su1)

He suggested caulis had bankrupted many farmers because:

> it's just a juggling game and then even with all the planning in the world you don't really get it right, like you know. Nature has a habit of dictating, at the end of the day.

In horticulture you plan, consider all the factors, make your best judgements and then hope nature dictates in your favour.

Growers are also at the mercy of customers, so they seek to balance plants' needs with those of buyers wanting uniform cauliflowers. Customer specifications and delivery schedules are crucial variables a crop manager monitors and aligns. Labour is the largest part of production costs, so a key factor. Take radishes:

> It's a fairly cheap crop to grow. The cost is in the labour of picking. And the time it takes to pick. But overall, it's an easy crop, its margin's okay, it's quite fiddly, you know it's not, not the fastest to pick.
>
> (L1.7)

This manager showed me a crop which had suffered extreme fluctuations in temperature and moisture, resulting in blemishes which would not pass quality control. He said probably it would be so slow to sort that it would be ploughed in; the cost of selling would exceed the cost of wasting it. Such situations are considered in yield forecasting, a fundamental task sitting at the intersection of people- and plant-centred variables.

Timeliness, timing and predicting the future

Many industries involve multiple complex variables, as growers sometimes mentioned when describing similarities to other jobs. But plants' perishability makes a difference; the need for fruit and vegetables to be fresh determines much in horticulture (Freidberg 2009). The pace and rhythm of plant cycles must coincide with human time (Brice 2021), which complicates horticultural workers' schedules as personal and supply chain deadlines do not always align (Medland 2021). Timings are tight:

> the supermarket will go 'well we want this much and we want this much', and they'll get a weekly one, then they'll do a daily one. So at eight o'clock last night they will have sent over their system how many tons they want

> from us today and the order was going out at one o'clock. And we'll either have some of it in the fridge, or we'll pick it this morning to put it in the fridge, for it to be packed and be sent away this afternoon.
>
> (L3.1)

One farm refers to it as "head on the block time" when they tell the packing house what will be delivered that week, because if it does not match forecasts, they can't fulfil customer orders. They constantly monitor crops, updating "guesstimates" of yield and picking schedule, monitoring numerous variables to refine forecasting:

- How accurately can you predict how much you're going to have of something ready to pick?
- 80%, 75, something like that. Is that a fair comment? What would you say? (L1.2)
- Oh, in terms of availability scoping, I say that they are pretty, pretty good at it now (L1.1).
- Well only because the suffering is extreme if I get it wrong!
- Yeah, because it, it goes back to the importance of it is so high and – but it's actually quite an art to it. But you also need time and resources and be able to comfortably go out there and look at it and, and know, have the right facts and figures, like how many were planted, but then you need the knowledge and experience to do it. I mean, some farmers are awful at it. And it really is a real – they'll completely misjudge it. So I think we're, we're pretty – we weren't always this good. You know, but I think over the last four or five years we've really excelled ourselves now.
- It does take a lot, you know, to work out what to do, and having done it a few times on certain crops helps, because you know, it gives you more confidence; you know how they're gonna come forwards, let's say. Not that anything's changed, you're just more familiar with it, so you can know how to react; you can be more confident in one respect.

One of this team sat me down in the farm office and pulled up two spreadsheets. The first was cropping data for crops he manages, a sheet for each year since he started more than a decade ago. Each year had a column for every week from planting to season end, with notes of anything significant such as extreme weather or when planting was delayed because a tunnel needed recovering. He logged the weight harvested weekly, then calculated the average yield per plant to forecast annual productivity, so the pack house knows how much to sell. He refines forecasts across the season, and adjusts sales as necessary, then notes actual yield against forecasts. His records compare varieties, and indicate tasks required each week, to forecast labour requirements – the "facts and figures" allowing accurate "guesstimates". The tomatoes are inspected closer to their predicted pick date to increase accuracy:

> a couple of days before I went round, and they'd just done a load of de-leafing on it, and then it was like, oh, there's a lot more orange ones than I thought. Which means in four days' time, there'll be a lot more red ones. So it's like, that's poor availability: it's double what I predicted. So that's part of it. I'll also be looking for, you know, potential problems or issues later on, so sort of looking a bit higher up [indicating further up the plant stem]. Have we still got lots going forward?.
>
> (L1.3)

He works from average yield, adjusting predictions based on observation:

> it's due to be really sunny, so I'll probably up it a little bit, but that tunnel there has got loads of whitefly in it, so I'll knock it back down a bit. Sort of juggling.
>
> (L1.3)

If you know how long it takes for orange tomatoes to ripen, you can predict when the crop will be ready to pick, and if you know the average yield of the variety, you can estimate what weight will be picked.

Smaller operations have less data for forecasting because it is not necessary or the grower lacks the capacity to gather it. More than one small grower told me they ended up in the industry partly from dislike of desk work, or struggles with academic skills, so working with data does not come easily. Other barriers are simple yet significant, such as the difficulty of using pen and paper in a field where everything gets muddy or wet, and there is nowhere to keep it. Growers hold information in their heads or basic notes, but still monitor crops, noting variables and inputs or timing, using experience to refine guesstimates.

Knowledge of plants, adjusted to conditions and interpreted considering previous experience, allows fairly accurate forecasts which are important for the crop manager's second spreadsheet: labour forecasts. His weekly breakdown estimates numbers required for tasks and picking intensity. A line is entered for each worker, so the annual recruitment profile resembles a diamond bulging around August's peak harvest. The previous year the recruitment manager miscalculated and so under-recruited, meaning some tasks were neglected, hence weedy leek fields and jaded weeders. Weeding had not been done in what he called a "timely manner":

> So you get jobs done in the little windows when you've got opportunity to do them. And, and you stay on top of things, you know? The plants like it, the people like it. The plants have no competition from anything else.
>
> (L1.2)

Weeds represent plant agency challenging the discipline of agricultural production (Argüelles and March 2022). Errors like under-recruitment are particularly acute because timeliness ensures freshness and regains some control over plants' growth. Experience and advanced knowledge of variables have refined horticultural production to increase the grower's control. Technology such as greenhouses intensifies producers' control of biological processes and nonhuman life (Moulton and Popke 2017). These developments culminate in controlled environment agriculture (CEA), which seeks to control variables affecting plant growth. Indoor, hydroponic systems reduce variation and unpredictability "as farming can get quite out of control if you don't have a … like, a nice tight grip on it" (CEA1.3). This indoor CEA grows salad and microgreens in a specialist aeroponic system regulated automatically to give plants what they need across the growth cycle. This "granular control" over inputs "can lead to higher growth rates and also healthier crops, essentially" (CEA1.2). The stability of a controlled environment means "all of the variables [to] meet" (CEA1.3), enabling consistency:

> You're also managing your environment. So your light controls, your – how much you're providing water. In our case, aerosol. You're also controlling the air temperature and humidity, different from day and night. You're controlling wind speed – so how much airflow you're providing your plants. High airflow can encourage growth by agitation but also can harden off your crop so you can basically make your crop stronger, which gives you a longer shelf life. You're also controlling CO_2 with parts per million. So your natural saturation of CO_2 in the air is 400. We like to enhance it to around about 800, so we double it, only during the day when the plants are feeding. And then once all of these little steps have been completed and you've managed to grow your crops from a seed to a very healthy looking plant, which meets QA, which is quality attribution – and that's normally … normally dictated by your buyer, how they want their crops.
>
> (CEA1.3)

There is less risk of variables going awry, meaning fewer judgments left to the grower who can monitor crops remotely via an app:

> you don't have to kind of interfere and be part of that process as much. You need less kind of like observation as well of crops.
>
> (CEA1.2)

CEA practitioners are often entrepreneurs or technicians rather than horticulturalists because "there actually isn't that much growing" (CEA1.1). But there are still risks, hence "vertical farming, 80% of it is cleaning" (CEA 1.1). I wasn't expecting a dishwasher to be essential equipment in CEA, given it is often presented as the future of food – high-tech urban production, highly automated, low staffing operations inspected by someone in a white coat.

In CEA the dirty fingernails of traditional horticulture are replaced by white coats and wellies: "You just want it to be super sterile just to keep out all the pests" (CEA1.1). Equipment is carefully washed down between crops to avoid contamination, the very opposite of the "muck and magic" tradition of organic growing which celebrates soil, and literally getting your hands dirty. This company doesn't sell produce, but generates income from developing specialist equipment and software. The crops are there to develop "recipes" for other growers to ensure a quality, predictable crop. The recipe holds the plant science information from a specialist who develops the details of sowing density, germination period, nutrient mix, frequency and concentration of aeroponic inputs which are fed automatically by pumps controlled by the company's app. A grower can buy the software, recipe and standard operating procedures (SOP) and has access to advice from engineers and growers. They do not need expertise in optimising growing conditions because they can buy it in the hardware, software, SOP and recipes. They still deploy skills of attending to plants during crop walks but technology contributes information, making data interpretation an essential skill. Materials and data are knowledgeable actors, and growers' skills change to enroll technology. This version of future food production requires different workers and skills, with fewer people required for harvesting and fewer "monotonous tasks", resulting in a "4.0 style food factory, let's call it" (CEA1.2). Crop workers will interact with the technical systems, using plant agronomy and observation skills, alongside robots which find it relatively easy to harvest from uniform trays in a closely controlled room.

The potential for robotic harvesting gives CEA control over something other growers find unruly. They have always navigated unpredictable "natural" variables, but now face another unpredictable condition: labour availability. This recruiter explained how the production team could no longer expect the workforce to respond at short notice:

> If we can forecast, you could say: 'right, we knew that today was gonna be sunny, we need 20 more people'. I can ring up the agents and go, 'can we have 20 more people next Thursday?', and they would be here Thursday morning, inducted and out picking in the afternoon. Now we can't do that.
>
> (L3.1)

Just-in-time recruitment which allowed picking orders for just-in-time supply chains no longer works, reducing growers' flexibility, making it more important to accurately forecast yields and labour.

Timing and timeliness are crucial to producing a good crop, and doing so profitably takes a grower "understanding the intricacies of planning, organisation and being ahead" (L4.2). This manager expressed disappointment that his most important expertise is organisational skills which address logistical challenges with staff and customers rather than agronomy. A field supervisor identified similar skills as key to her role:

> So it's a lot of time management – I guess – skills go into it and sort of forward thinking, especially because you're thinking of the next day and the next day after that, so you can plan what you need on the field because if you don't have something on the field then your whole team will be waiting and it's not good. Time is money, as they say, apparently.
>
> (L1.5)

She described a complex process of disaggregating orders into boxes to be picked and daily rates for each worker; numbers of people, plants, boxes, cropping and picking rates must be spot on. Her role is not managerial, but her abilities are central to timely operations because she supervises field-workers whose efficiency is paramount: "time is money". Their roles need further consideration because they are so often excluded from attention to skilled work.

Becoming a champion picker

> A Champion can be defined as one who achieves superb results (e.g. very high output), consistently (day in, day out), while meeting required standards (e.g. picking to supermarket specifications, not missing or leaving any ripe product) without needing constant supervision.
>
> As growers, you source the best plants you can, give them the best conditions you can and nurture them to keep them healthy and productive. We need to think of labour in the same way. New workers are raw materials and will only reach their potential with the very best treatment, including good facilities, fair treatment, pastoral care, thorough induction and training and proper remuneration.
>
> (AHDB 2018)

This idea of a champion picker was promoted by AHDB in response to recruitment struggles, aiming to maximise workers' productivity, focusing on soft fruit as the largest user of seasonal pickers. AHDB provided training and resources for coaching workers to become champions. A video titled "How to pick soft fruit: becoming a champion picker" talks novice workers through getting started, including health, safety and hygiene in multiple languages. Viewers are told the video will help them "become a better picker and earn much more money" through discovering "the techniques of the very best pickers – the champions". Basic picking techniques are demonstrated by pickers who show "the correct way to pick". The camera zooms onto a hand:

> To pick a strawberry specifically, pinch the stem gently with your thumb nail against your forefinger, and flick your wrist to snap the stem. If you squeeze too hard it will not snap.

There is an instruction to ask any time you are unsure and to pay attention to the supervisor: "Look and listen carefully as there is a lot to learn". The "right" methods are fast methods: "the best pickers get it right first time every time [...] This method is wrong because it is slower". Sped up footage of pickers working along strawberry tables plays whilst the narrator explains "To keep going at a high pace for up to eight hours, altering your posture as needed will reduce aches and pains".

Novices are advised to keep practising techniques until they get it right:

> Champion pickers have a good search pattern. They work systematically, whether this be by each strawberry plant or smaller sections of raspberry plants from top to bottom. They do not move their hands at random or go back to pick missed fruit as these are things that slow them down. You must develop a pattern to become efficient in not missing any fruit. They pick with both hands, not just their dominant hand. Start by picking with both hands but slowly to get used to it. And then when you become comfortable enough, speed up.

Good workers are fast workers, earning more money for themselves and the company. Inefficient movements are highlighted, reinforcing the most efficient ways of moving and organising tasks. But a champion requires more than good technique:

> What is your motivation for becoming a champion picker? Set this as your goal and keep it in the front of your mind each day on the farm. This will help when some days become tough due to hot weather. It will keep you motivated.
>
> Picking is hard work. It is physically tiring and can involve working in difficult conditions. Drink plenty of water, eat well and make sure you get plenty of rest.
>
> Pickers become champions by observing other champions. Make a point of finding the champions on your farm, watch how they work and copy their techniques.
>
> Champions set themselves a pace that they can maintain throughout the day. Other pickers may start too quickly and then slow down towards the end of the day. Champions keep going and because of this they earn more money. They stand out from the rest.

This advice reveals that pickers are expected to do arduous work requiring physical and mental resilience, implying endurance and tolerance are qualities of the best pickers.

I met a champion strawberry picker, who was earning around £19 per hour whilst the slowest worked at a rate equivalent to less than minimum wage, costing the company. They would receive a visit from an experienced supervisor who coaches slow pickers. His coaching is informed by analysis of the champion's technique, as the grower followed AHDB advice to film fast pickers and watch closely:

> It's a knack more than it is actually something that can be completely taught. It's really weird. I watched the videos and I looked at them, and I think it's really weird the way they do. And you can try and copy, and everyone can copy, but it just doesn't come up – it doesn't flow. It's just – I suppose it's – I can't find the right analogy to do it – riding a bike? Some people can ride a bike really effectively from day one, some take 20 years to do it. There is just something in the way that they do it that is just – and we don't know if they've got wrists that are slightly dislocated – their eye-hand coordination, which is really high. Eye-hand coordination is really important because they'll be looking each – when we videotaped, what they do is, look for the next one before they've picked the one that they've already got. So it could be just something as simple as that. And obviously, not everyone has got as acute high eye-hand coordination as they've got.
>
> (L3.1)

This conveys a sense of wonder or incredulity that anyone can be so fast, then attributes it to some innate "knack" or physical disfigurement. It denies the potential to acquire picking skills through learning and practice – they just *happen* to have acute eye-hand coordination.

Contrast with analysis by the consultant who supported this business to study picker efficiency:

> It's really interesting actually 'cause the perception of the industry was it was down to hand speed. So how quick you move your hands, that's the thing that determines and if you'd asked 'what's the answer to the problem?' it's 'you need to get people to move their hands quicker'. When you actually look at the videos, it's actually the preparation that the individual does. It's how they organise themselves. So it's not – the hand speed is more or less the same. It's how they organise themselves. So it may be that you organise them so they've got all the punnets around and therefore the movement of the hands is reduced in terms of distance travelled. It was just simple little things like that, that you'd never see with a naked eye.
>
> (Sh2)

The Champion Picker video highlights efficient set-up and handling of materials, recommending keeping empty punnets near the body by swapping full ones to the back of the trolley with one hand whilst the other sorts fruit. These movements are so fast and smooth that the efficiency in working with materials is easily overlooked, but the skill is not innate to their body, the technique was learnt, practised and refined.

Watching the champion picker at work showed her logistical awareness to be key. My guide explained how the consultant helped them notice she nudged the trolley along with her hip so she could keep moving continuously without pausing to look down or push the trolley by hand. The polytunnels are on a slope, meaning that when working downhill, pickers struggle to control the trolley, so she had taken a piece of baler twine, tied one end to the trolley, the other to her belt loop, allowing her to bump it along without it freewheeling too far. This ingenuity – not any double joints or extraordinary physicality – made her a champion. Nor had she brought a "knack" to the job; she had observed, judged and responded.

Speed is not everything as pickers must monitor quality, and select produce meeting specified standards. Having been shown what a punnet of the correct weight looks like, an experienced picker will "get their eye in" to select the right fruit whilst maintaining speed:

> I guess you have to be quite quick making judgements because you don't really have time to observe every little leaf. You have to be quite quick at judging the crop as you pick and finding the quickest ways and most efficient ways.
>
> (L1.6)

One fieldworker described how their team spent a day working out the fastest way to pick leeks, comparing techniques, then landing on gripping the shaft in the optimum way before pulling. These efficiencies make a huge difference when scaled up, especially for low margin crops, making experienced returnee workers invaluable. This was apparent during 2020 when the pandemic prevented many seasonal workers from returning to their usual workplace, prompting unprecedented levels of domestic recruitment. The impacts were clear:

> We didn't get some of the more experienced people early in the season that we would like, you know, so we were missing experience. 'Cos we had a lot of non-returners last year, so the people that came needed more training, and we didn't have as many experienced supervisors to train them. So you know, we were up against it, we had productivity issues.
>
> (L2.1)

Why local recruits were less productive is unclear as employers' assessment drew on racialised narratives of British laziness versus Eastern European drive and rigour. Some wanted to disprove the suitability of local recruitment to support their case for an extended seasonal worker scheme. But experienced returnees bring efficiency as "they'll fall into the routine" (L3.1); their know-how enables them to get up to speed quickly, bringing productivity which increases profitability.

How pickers refine their techniques and are trained to increase productivity demonstrates that seasonal work entails learning and acquiring skills. Picking accurately at speed is not an innate ability, but a practised technique; champions are not born but made. Their skill extends beyond the individual champion, as they enroll twine, move punnets to gain speed, then share expertise with colleagues. Watching experienced pickers at work, their speed and focus appears effortless, moving without seeming to pause or search for the fruit as their hands deftly find it, check and sort it into punnets. Their bodily intelligence appears so smooth it is easily neglected (Patchett and Mann 2018). The task seems monotonous, but they respond to individual fruits, mentally comparing them with the spec sheet, rejecting those off-spec. Skilled pickers combine judgement, speed and dexterity, assessing the specific context and adjusting their movements in response, demonstrating the facets of skilled practice (Ingold 2000). Although repeated many times, the task is never exactly the same, so motions are not repeated mechanically, rather the skilled worker moves in response to materials and goals, at speeds and for durations required in industrialised contexts (Carr and Gibson 2016; Klocker et al. 2019). Staying motivated whilst repeating the task is another attribute of a skilled picker.

Figure 5.1 A champion picker harvesting strawberries.

A "knack" for efficient picking: Underestimating workers' skill

One recruiter reflected on efforts to attract sufficient pickers: "As much as they say it isn't skilled work, it is". But even those working closely with field teams underestimate their workers' skills, as apparent in the previous section. Some struggled to align their first-hand appreciation of the work's complexity, with social imaginaries of these jobs as unskilled; notions of manual tasks as unthinking often reigned. This tension was encapsulated by a manager telling me most tasks done by fieldworkers are not skilled:

> At the end of the day, a lot of it is a knack rather than a skill, if you see what I mean. I couldn't quite describe what the difference is with that.
>
> (L1.3)

I pushed him to differentiate between knack and skill:

> Yeah, I suppose a lot of it is very repetitive, I think. That's probably what it is. So a lot of it, you need to do it right – so again, doing the side shooting of the tomatoes it's something that with time you start off doing it, and then you start off being really nervous, and really tentative because I've said 'don't kill the tomato plants' and they start off really tentative. But you'll do that every day for probably two months, or for part of the day. You know, you'll do a bit of maintenance. You'll do your patch and then you'll go and side shoot cucumbers and side shoot tomatoes so after a while repetition just builds up that sort of ability, sort of thing, and it ends up being second nature, whereas I suppose when I differentiate it's a skill it would be perhaps, I don't know, rotavating with a tractor to a particular depth sort of thing which is a more technical thing and some people, no matter how many times they go over a field, won't get it right and some will. Whereas with tomatoes most people, if you do it often enough, will get it right. It's the repetitive nature of it and it becomes natural and you pick it up after a bit just because you've been doing it so often.
>
> (L1.3)

This explanation reveals assumptions about what counts as skill – technical, specialist and learnt – whilst technique acquired through repetition is perceived as a lesser form of ability. It implies mind-body dualism typical to accounts of manual work as unskilled and neglects thinking done by bodies in action according to an ecological definition of skill. He sees the "knack" of picking or side-shooting centred on repetition of movement, like a habit so quick it is easily under-appreciated, overlooking work of observing specifics and responding, exactly what Ingold defines as skill. His view indicated that "they" who say it isn't skilled work includes people within horticulture, who undervalue skilled pickers' know-how.

The counter view was expressed during a tour of a mushroom growing operation resembling a factory. Dressed in white coats, overshoes and hats, we looked at the long peat beds stacked to the ceiling, and my guide explained how a picker has a section where they monitor each flush of mushrooms. They pick to match the specification and to influence the next flush, for example, removing one from the centre of a clump to allow others to grow larger. A good picker manages the first pick of button mushrooms to ensure later flushes of larger flat mushrooms, observing and selecting to influence growth patterns. My guide described this as "creating" and "controlling" their crop, terms suggesting agency, intelligence and judgment appropriate to their assertion that picking is skilled work. The mushroom factory would easily be assumed a deskilled workplace where jobs lack creativity, self-determination or tacit knowledge (Timmermann and Felix 2015). But the pickers orchestrate what happens on their bed, becoming proficient in managing its flushes and moving as an efficient picker.

To learn the bodily capacities of mushroom picking, a new picker is buddied with a more experienced colleague who shows the technique, initially one mushroom at a time, practising towards cutting several at once. Speed increases until picking seems effortless, as the picker develops "tempo" – a rhythm at a nice steady pace, not flat out, allowing time to judge each mushroom, but flowing. The description of tempo was as poetic as the motion, with my guide likening it to a dance. Appearing smooth and easy suggests it is an unthinking automatic habit, proficiency which easily makes skills disappear (Attewell 1990). But pickers demonstrate constant interaction between body-brain-environment in skills of thinking-acting-responding (Patchett and Mann 2018). Work like mushroom picking has been falsely regarded as deskilled, but attending to their practice makes skill apparent.

Ease gained through repetition was described by another manager as "getting the hang of it":

> we would expect them to progressively get faster as they get more used to it and how to do it, and then it becomes second nature.
>
> (L1.2)

His colleagues had also described pickers' work becoming "second nature" when explaining what he meant by a knack, suggesting something habitual or instinctive, automatic action, rather than thoughtful consideration, routinised physical abilities rather than mental exertion. Pickers are dismissed as having knack not skill, or assumed to be freakishly double jointed rather than talented, because they make it appear effortless, and because bodily intelligence is readily demeaned. Dismissing ability as bodily habit has repercussions: it enables discourses and control systems which categorise some people – typically of marginalised gender and race – as unskilled, in spite of their expertise. Second, it aids the illusion that skill can exist apart from what bodies do, reinforcing notions of it as a free-floating capital, rather than as situated practice which lives through people moving, learning and doing (Iskander 2021).

A tendency to underestimate bodily skill echoes through descriptions of strawberry packing workers who adjust the weight of punnets rejected by the automatic line, adding or removing the fruit. One of their managers marvelled at the scene:

> I'm absolutely fascinated watching the girls. The punnets come in and they literally are just looking at the scales, doing this and moving them across and they're just watching the green lights to come on. And when you watch some of them you just wonder how the hell they do it.
>
> (L3.1)

They do not appear to look at the fruit, judging by feel, moving it almost by magic. Note the reference to "girls" on the line: perhaps conscious that he was describing colleagues as "girls" to a woman, he went on to clarify that it is mostly women working on the packing line. Men put on the task are much slower: "women seem to be able to be more dexterous than men". Gendering of delicate, speedy horticultural tasks is common (de Castro et al. 2020; Nieto 2014), and naturalises workers' abilities, suggesting they are an innate capacity rather than learnt and practised (Osterman et al. 2022). Abilities regarded as "just raw corporality" (Iskander 2021, 15) are not skill, and hence these remain unskilled workers. It's knack, not skill.

Similar assumptions were apparent in the large organic grower's attempts to recruit people with suitable dexterity, suggesting some people have the right body for the work. The selection process involves a short time in the field, so a supervisor can demonstrate tasks, then observe their efforts:

> with such a small group, it's really easy to see the ones that just get it straightaway and the ones that struggle a bit more [...] Some people just don't have – they're not very dexterous. It shows you pretty quickly who could be cut out for this work. They're not going to do it perfectly but you get a pretty good idea of someone who just can't do it at all.
>
> (L1.4)

The supervisor identifies who seems to lack hand-eye coordination and recommends they are not hired. A combination of dexterity and focus sets apart those suited to the role:

> Some of them are just brilliant. They just really concentrate on what they're doing, looking at what they're doing and trying to be precise and you can just see in people the way they hold things. Some people just can't hold things right, just can't cut.
>
> (L1.4)

Good horticultural workers need dexterity and concentration, abilities traversing the typical separation of mind and body, indicating that work assumed to be mindless engages mental aptitude.

In fact, the most noted quality of good harvest workers was "the right attitude", meaning being conscientious and willing: "It's about attitude, it's about wanting to be here and wanting to do it" (L3.1). The "right attitude" means hard-working, enthusiastic, motivated and efficient – a productive employee:

> People with the right attitude, a good working attitude, reliable, industrious, you know. That's about it, really. You know, we need people to turn up. A, turn up, B, work well during the day and that's it. Some people aren't reliable and aren't very industrious.
>
> (L1.2)

Like dexterity, these qualities are regarded as innate:

> They've got to have it in them to start with. They've got to have it in them and then they can see a future and … Do you know what I mean? They've got to want to do it. And you can soon spot them in this business, you know.
>
> (L4.1)

This imagines good horticultural workers are born not made, and that most often they are not born of White British heritage:

> The British people we have had into the packhouse have tended to come from traveler backgrounds, that have stuck it out, so their work ethic is different.
>
> (L2.2)

Bulgarian and Romanian people who currently dominate UK seasonal work are perceived to be motivated, hard-working and responsive to financial incentives such as piece-work. Good migrants are imagined as born not made, replicating the racialisation of "hardworking bodies" common across industrialised horticulture. There is another trope recurrent in descriptions of fieldworkers' bodies.

Back-breaking work

Across sites and organisations, I often heard horticultural work described as back-breaking. This expression was not always meant literally, being a common description of strenuous work, but conveyed the contortions and strains horticultural bodies go through. Conditions in the UK may not be as harsh as those where less regulated farmwork is dangerously unhealthy (Holmes 2013), but bending, lifting, twisting, squatting, reaching, again and again, literally damages backs. As part of his graduate training scheme, this employee spent time harvesting salad alongside fieldworkers for a couple of days:

> I've done some pretty tough and rubbish, sort of jobs, previously. But I think that was very much up there, like, you know, it was back-breaking. And I think, you know, it's a long, long, very long season, long, long days, long, long weeks, long season, and, yeah, that's before you consider doing night shifts, and things like that. And, you know, working a long way from home for a lot of, well, for the majority of our colleagues there, so, yeah, no, it's very tough.
>
> (L2.3)

He thought it important to experience this work which comprises the "beating heart of the business" to appreciate the physical demands. But back-breaking work presents a tricky balance for his HR colleagues who don't want to deter recruits whilst needing them to have realistic expectations:

> We're really honest about that, really honest about how hard the work is so the fact that you are gonna be bending for many hours of the day, you know, it is gonna be a lot of kind of – your back is gonna ache, your knees are gonna ache. You know, it's bending down, cutting the lettuce, making sure it's good quality, like there's lots of different areas of it. You're gonna be in like the sun, like we've got canopies and things but it was a hot summer last year, like you were gonna get very hot.
>
> (L2.4)

Recruiters look for signs someone is physically tough enough to endure outdoor work in the UK. Good workers are hardy and ideally arrive as bodies already holding qualities of endurance.

Endurance is also required for managerial roles because of long hours and constant pressure as this manager described:

> So it's being able to take that burden and pressure and carry on where necessary, I'm not saying have a nervous breakdown, you know what I mean. Be able to just, be able to have a normal family life and be able to socialise and do everything but you have to have a very busy life. You learn to probably get up at four or five o'clock in the morning, get lots of stuff out the way, might get home and see your family at six o'clock because there's something important going on for two hours and then at 10/11 o'clock be on your laptop sorting something out for the following day so you have to sort of live that sort of lifestyle and family probably doesn't come first, probably wouldn't say that to my wife but I think she knows it, and it's you know, the job pretty much comes – the job has to come first if that makes sense.
>
> (L4.2)

I asked why he was willing to endure this: "it is a way of life and if you're brought up to that way of life". Farming is the "lifestyle" he was raised to; good growers seem born, not made.

Whilst talking to a field supervisor at the large organic farm, I noticed she occasionally put a hand to her lower back, bending back and forth. I asked if she was struggling: "yes I am actually". A colleague had just given her a massage to try and ease it; she had also taken painkillers: "so that should just sort it out. Got my holiday next week. Just got to push through" (L1.5). Stopping to chat seemed a much-needed break before she rejoined the team bending over between the leek rows to weed. We later spoke about the relative demands of different crops and I was surprised to learn the team's least favourite, expecting something heavy, dirty, low to the ground, physically demanding to pick. But it was courgettes, a plant I regarded as pleasant and light. Her team knew well that courgette leaves are prickly and an irritant, whilst fast growing fruits need regular picking. Harvest workers require physical endurance to do their jobs and feel precarity in their bodies (Stead 2021). That season the pickers were given protective sleeves designed for cyclists to prevent rashes. The person responsible listens to the team's feedback, then tries to provide "anything that might make their lives easier", a lighter knife which creates less strain, knee pads, good quality waterproofs. Where employers are less attentive to workers' comfort, they find their own solutions. I saw one team picking French beans had tied bin bags to their belts so they could carry a good amount before needing to walk to the cart or move it along the row. Fieldworkers demonstrate ingenuity and adaptability to become faster and more resilient, suggesting they have the agency to reshape work conditions (Binford 2012; Gertel and Sippel 2014; Rogaly and Qureshi 2017).

A good grower

During an IPM training day, I joined growers and specialist consultants touring two farms. Growers were guided to signs of aphids or thrips on the plants and were soon inspecting foliage, spotting damaged fruit or traces of insects, sometimes calling the specialists over to identify them. At the end of the visit, one of the consultants summarised the principles and benefits of IPM, adding with a smile "All growers have to be entomologists". After a pause, he expanded, adding that growers also need to be soil scientists, and agronomists, and … letting the list tail off. Growers need a range of expertise, and although they have rarely had training equivalent to a degree in entomology, the awareness and skills they develop are attuned to monitoring plants. But as the quote opening this chapter highlighted, a good grower works at the conjunction of inward focus on the plant and outward perspective on the world. They need a complex array of knowledge and abilities:

> I'd say it's someone who's physically very hardworking and willing to put in a lot of hours in all weathers. And is – and can basically take on all the

> basic elements, understand all the basic elements of rotations and crop planning, fertility building and also then, well sowing, planting, weeding, harvesting and actually the, the, kind of, marketing what to do with it all at the end.
>
> (S2)

It goes beyond plant work:

> It's having a variety of skills, so it's not just the growing. You're sort of – yes, you need to be a good grower, but you've also got to manage people and be patient and … yeah. It's, it's quite – can be quite stressful, but you probably need to be fairly relaxed about the job and sort of go with the changes, 'cos they move so quickly. You can sort of set your mind thinking 'this is how it's gonna work' but two weeks down the line that plan's gone completely out the window and you have to start again 'cos you're so reliant on the weather.
>
> (L3.2)

They need to attend to all those variables, including workers' availability and productivity, and "cut the right path through them" (L1.2). Efficiency needs all aspects of production properly coordinated. The spring onion picker's speed depended on others ensuring correct soil moisture, and there were only onions to pick because everyone across the production process had coordinated seed with the eight essential inputs for plant growth, coaxing the onion's skill in growing. The skills of growing food extend across and between everyone in an operation, including materials and plants, making coordination of each actor's movements an essential horticultural skill. Skilled growers do not just attend to weather and leaves, but movements in finances and policies; moving numbers on spreadsheets and people into vacancies are as important as swinging tools into soil, each demonstrating the judgement, care and dexterity of skilled work (Pitt 2013).

This complex array of activity is managed in a very pressured context, sometimes by one individual, so:

> You just have to be adaptable and I think adaptability is an important thing, being able to understand that and go with it.
>
> (S1.1)

In addition to – or perhaps across – the technical abilities a grower needs are capacities for flexibility, being ready to adapt when nature dictates, applying skills of attending to change in the array of variables affecting plants. Attention and adaptability are features of plant work being never wholly knowable or within a person's control (Pitt 2014). The limits of horticultural knowledge were highlighted by individuals at various points in their careers, the first a recent graduate who placed himself at the middle of my scale of horticultural knowledge:

> I think I know the very basics of what you would need to know to get started but you know, I still have conversations with growers and and there's things that I've never come across three years in. And I imagine there still will be in, if you ask me again, in 15 years' time, I'd still probably say I was at a five. Just because I think there's always things that you don't know. And you, and you never know what you don't know. [...] how do you ever know that you're doing it as best as you can?.
>
> (L2.5)

This could be modesty reflecting junior status within the company, or a scientist's epistemological reflection on the impossibility of absolute knowledge. But one of his colleagues suggested even the most experienced person in the company would not claim complete knowledge:

> if you spoke to our head grower, [and] probably one of the best farmers in the country really, you know, and he's in his 21st season and he, almost every day, every weather event, every field, every variety, you're getting different challenges, so I mean, even he probably wouldn't say he's, you know, anywhere near a 10.
>
> (L2.3)

A similar view from a lifelong farmer suggests accepting ignorance is part of horticulture:

> I'd probably have given myself an eight 10 years ago and I'd give myself probably a four today. Because the more you're in it the less you really realise you know. And you know, you always have seasons – and I've done it – where I've absolutely got it sorted, it's been wonderful, amazing and you just go 'wow' and then the following season you get such a kicking you realise you know nothing.
>
> (L4.2)

No matter how good the grower, "you're in control of very few aspects of your role" (L4.2). Lack of control suggests horticultural work has been hugely underestimated, as directing crop growth requires sophisticated intelligence including insight into vegetal lives – the exact opposite of how plant work has typically been regarded.

Conclusion

This chapter has shown the breadth and depth of skills in horticulture. And they say it's unskilled work. Detailing this work was important because it is so often assumed to be unskilled. This account expounded my ecological definition of skill, demonstrating how growers attend to the environment then move others into place to encourage plant growth. I emphasised the bodily intelligence of skilled pickers because their work demonstrates the fallacy of categorising work as either intellectual or manual, or dismissing it as unskilled. We encountered numerous processes and roles which must align to get fresh produce to consumers on time, indicating how skill extends across people and things. Although often regarded in ways which individualise it, skill is a social practice and collective achievement. Coordinating this complexity, aligning inward and outward views from plant to world, is a fundamental horticultural skill. But plants and their interactions with people are never wholly controllable or knowable, making skills which plant producers depend on unlike those in other industries.

Perhaps surprisingly, we have seen how some within horticulture reproduce narratives characterising horticultural work as unskilled, regarding tasks executed with speed and aptitude as achievements of innate physical traits rather than learnt, practised skills. Such depictions can reproduce gendered, racialised assumptions about types of bodies associated with these abilities; they also individualise skill, locating it within certain bodies. This reveals the extent and depth of mind-body dualism underpinning understandings of work and skill, which render "the knack" of being a good worker a capacity inherent to a body rather than recognising intelligence developed through experience. To understand how these conceptions are perpetuated, it is necessary to consider how people learn to become good growers, through often overlooked learning and skill development.

6 Becoming a good grower

Introduction

The previous chapter outlined the complexity and breadth of skills required to be a good grower, so how do people acquire this know-how? Very few of the growers I met hold a professional qualification in horticulture; several had degrees in agriculture, but their formal education included minimal or no specialism in growing. This reflects the limited availability of courses designed to prepare people for a career in fruit and vegetable production and typical pathways into these careers. Learning on the job dominates, with knowledge exchanges between peers and within workplaces significant for skill development. Across all stages of experience, the key modes of worker learning are through showing and doing. By examining these learning processes, it becomes apparent that these skill pathways and how they are understood can undermine the value of horticultural knowledge and work. It is also clear that a good grower is never self-made, as they rely on others with growing skills, and flows of knowledge around horticultural communities.

Show-how know-how: learning by doing

Showing a recruit to horticulture how to do their job begins at induction, an extensive task for large recruiters. One uses a tailor-made animation which starts with health, safety and welfare issues, "so nobody has any excuse because it's all pictorial" (L2.2). After these necessities, instruction focuses on doing the job, mostly through pictorial aids such as "spec" sheets specifying what to pick. Visual communication overcomes language barriers, and trainers suggested many workers prefer it to reading as they are not very "academic". Face-to-face instruction also centres on showing as more experienced workers demonstrate the task or technique, and the inductee observes and then mimics. Co-worker training exists across horticulture (Hernández Romero 2012; Klocker et al. 2019; Pye Tait Consulting 2020), reflecting its popularity as simple, cheap workforce development (Lam and Marsden 2017).

DOI: 10.4324/9781003281689-6

There are also popular beliefs about the best way to learn horticulture, which prioritise direct contact between novice and expert, making peer learning significant in the industry.

One manager summarised his training approach as: "telling and then showing and then checking, I suppose, is probably the three things I do" (L1.3). The worker is shown, then observed and corrected. Showing how to do a task is about demonstrating an efficient, safe technique and communicating the context so workers understand why doing it right matters. To explain how far apart tomatoes need to be planted, the manager brings a trolley to show that rows need to be wide enough to allow pickers to pull it through. The supervisor similarly demonstrates the importance of picking to spec:

> I can show them quite quickly if it's not being done correctly, what would then happen to that produce if it went into the crate, into the packhouse, got taken out, goes into the box, how it would end up looking and to explain to them why it's so important that we do things in a certain way in the field to preserve them, to get stuff in there in really good quality.
>
> (L1.4)

Those responsible for training teams value visual learning and rarely rely on verbal instruction alone. They have little faith in trying to communicate remotely from the site and practices where workers will be immersed:

> I personally think the 'showing people how' type thing is really, you know, is an excellent way of – you can't beat it really [...] the guy showing you how to pick a leek, you know, it's right in front of you. That's the best way. I don't know – I don't really know how else you can – I mean, how else could you do it? I mean, how would you show someone to pick a leaf? [...] It's just not good enough, like sort of giving out kind of abstract – barking out abstract orders for sort of relatively complex things.
>
> (L1.1)

This comment highlights that supposedly simple horticultural tasks are actually complex, whilst suggesting understanding them as tacit knowledge not easily verbalised or explained (Gertler 2003; Polanyi 1958). The "knack" is more bodily than intellectual and so not readily put into words or communicated. What is not clear is whether this is because the knowledge itself defies explanation, or that instructors are not equipped with skills to translate embodied knowledge into instructions. It is likely a combination of the two, given that expert practitioners typically instruct novices by combining doing and talking about it (Lave and Wenger 1991); even bodily know-how can be described or spoken of (Hitchings 2012; Ingold 2000).

Learning by doing is the second key mode of skill development in horticulture; show-how and know-how interact as workers practise and repeat tasks. Typically a supervisor or buddy spends a period alongside new workers:

> so normally I'd have someone completely new to the job with one of the boys who's been doing it for a while, who knows what they're doing and they'd show them what they're looking out for, so what they look out for quality wise and size wise.
>
> (L4.3)

Working side by side allows showing and doing to combine in workers' learning:

> So what it says on that spec sheet is how we've got to pick it so you're just given a demonstration of that and then you let them have a go and you just kind of look at their technique and stuff and then I'll just assist them: 'okay, instead of holding it this way you hold it this way, grab it with this hand' or whatever it is. Then kind of just leave them to it for a while and then sort of after about two hours usually check in with them, how are they going, if they feel like they need any extra help or any extra tips or whatever that may be and then just explaining what else needs to be done throughout the day.
>
> (L1.5)

The trainee starts by watching, then is watched as they seek to mimic what they were shown, with the more experienced colleague correcting mistakes. I asked a volunteer at a small organic farm to describe how he learnt a task like the weeding we were working on:

> In that case [paid worker] guided me, so, I'm guessing, [farmer] had taught her, but she would very specifically, literally, with, like, hand in the soil direct me what to – or just say what to look for, what to worry about and what not to worry about. And, actually, just, sort of, show me two or three examples and then let me go for a while. So, very hands on.
>
> (S1.4)

In this combination of showing and doing, the volunteer was shown, then did it themself and then repeated the actions: "I'll just copy and paste until its finished". Showing quickly transitions into doing, as the novice moves closer to the centre of the community of practice (Lave and Wenger 1991).

Given the depleted infrastructure of UK horticultural training and education, reliance on experiential learning may have arisen through necessity. Industry bodies suggest businesses struggle to find provision which meets their workers' needs (Pye Tait Consulting 2020). But what growers convey about the skills of being a good grower suggests that learning by doing is a feature of horticultural know-how. If they are to adapt to ever-changing, complex conditions and weigh up multiple variables, knowledge learned by rote is inadequate. Conditions are never exactly the same, and growers must know how to respond in each case. The idea that learning best entails experiential dimensions was suggested by one of the more specialist trainers I spoke to – a consultant who supports businesses with LEAN. He described his preferred training model:

> So as soon as we cover an element of theory, it's 'now go away and actually apply it and let us know how you go on with it'. So, you know, I think people feel like they're using it and they're getting the benefits in the early days.
>
> (Sh2)

This motivates trainees to adopt LEAN principles as they soon notice practical improvement. It is also a typical learning by doing strategy for skills development, with prior knowledge passed on and combined with problem-solving to develop new knowledge (Lam and Marsden 2017). Even in larger commercial operations and for those with agricultural degrees, background knowledge or general principles are supplemented by learning on the job. I encountered two justifications for this: in one company the director likes to "mold people" to their ways, so avoids bringing in people with habits from elsewhere. Otherwise, it was described as inevitable because the level of detail specific to a production system cannot be taught in relatively generic degree courses.

Learning by doing has gained particular credence within agro-ecological growing communities (MacAuley and Niewolny 2016; Timmermann and Felix 2015). This is regarded as a knowledge intensive mode of food production, which can't be learnt as prescriptions or codified information because of place-specific practices (Carlisle et al. 2019). Learning by doing is central to how people become agro-ecological growers, with beginner or new entrant growers often doing several placements or jobs – including volunteering – to gain experience (Claeys and Van Dyck 2022; Ekers et al. 2016; MacAuley and Niewolny 2016; Taherzadeh 2020; Weiler et al. 2016). This has arisen at least in part because of shortages of formal educational pathways in agro-ecological learning. This farmer initially sought a degree to transition from their previous career into agriculture, but:

> There were two frustrating things. One of them was the realisation immediately was that actually it was an organic module which was tied onto a conventional agriculture course and b), I was going to spend the next two years in a classroom. And I'd come out probably trained as an agronomist having a real understanding but actually no soil time. Or the alternative was to go – if I could have found one, because there are still a few around – but go to a straightforward agricultural college. There aren't many of them left, not like when I was growing up in the seventies. Again,

> you'd either be in horticulture but really utility horticulture, looking after ground in a very conventional way, or driving a sprayer safely. None of these things ticked any of the boxes so I really didn't know how to learn and there really weren't any alternatives that I could access. So I just had to go and do it.
>
> (S1.1)

There followed a series of jobs on farms, progressively taking more responsibility:

> Yeah, I learnt on the job. Yeah, I read all the books; I continue to read all the books and then I sort of taught it to myself and I learnt on the job. I was very lucky that when I went to a farm in Hampshire, I'd been there in position for about 18 months and I was growing vegetables as an assistant and then I was thrown in really at the deep end. And I had the opportunity to learn on someone else's dollar, if you like, you know. And that was very – I got to make some big decisions very early on. I had realised I was quite good at it and I realised that the person I was taking over from, [who] wasn't actually very good at it at all!.
>
> (S1.1)

Such pathways are common amongst organic and small-scale growers, who supplement classroom education with hands-on experience. Growers elsewhere have been found to prioritise hands-on learning over theoretical, class teaching, but this may be as much about status and identity as acquiring embodied, place-based knowledge (Feinberg 2020). This farmer's history demonstrates the significance of having the capability to access skills development, through relevant learning opportunities and growing without risk – "on someone's dollar". The ecosystem around the learner matters; access to these abilities is not equal and depends on the resource infrastructure or impediments to knowledge flows (Pitt 2021b). Attention to this context and how it impedes learning is the focus of the Chapter 7.

Learning by doing is associated with experiential knowledge, practical or manual work, which often implies embodied knowledge (Rose 2005). Experiential learning's dominance within horticulture therefore feeds the notion that this work is more bodily than intellectual or mindly – knacks not skills. But within horticulture, the hierarchy of knowledge is sometimes flipped, in a tendency to valorise that which has been learnt experientially: literally "having your hands in the dirt" (S1.4) is regarded as vital. One field supervisor was dismissive of those without practical experience of her role:

> some of the managers they've never picked anything a day in their life so it's very interesting how they can kind of dictate to you how something should be picked if they've never done it themselves.
>
> (L1.5)

An office-based superior can't instruct fieldworkers, because they lack appropriate knowledge, and "because it's just about finding – finding your way of doing it" (L1.6). In this case, it was working out the most effective way to pull up leeks, which the team did individually and then through discussing their method and comparing techniques. Experimentation and improvisation were part of their learning, combining showing, telling and doing, but this required a degree of autonomy not all workplaces offer.

In celebrating hands-on learning, growers expressed ideas of knowledge stratification, a widespread perception that some kinds – codified, standardised, intellectual – suit classrooms, whilst experiential learning is necessary for tacit, practical know-how. These distinctions rely on attempts to separate what are overlapping and often co-located (Allen 2002; Gertler 2003). Many within horticulture fall back on these assumptions, but in practice, their learning demonstrates interconnected knowledge flows. One might assume technical specialists such as agronomists and entomologists rely on formal education centred on information in texts and generic principles, but they also need experiential learning. One consultant told me many of their recruits enter with a general science or environmental degree, then train in a specialism on the job, building on standardised knowledge through experience. I asked how she, after years as an agronomist advising growers, knows what is happening with a crop:

> for my FACTS[1] training ['cause] we'll get shown examples and taught examples and what a typical nutritional physiological disorder is. And the other one's just experience and colleagues who've seen it before. So if we get something really odd, we'll quite often just send an email and say 'Has anyone seen this before?' And then if not, we go into investigating, sending things for leaf analysis and even sometimes looking for a pest.
>
> (Sh2)

Like growers, technical specialists interpret a specific context, relate it to their general awareness of plants and insects, and make a judgement informed by prior experience. Across agricultural practices, decision-making combines formalised and tacit knowledge, judging what this specific plant is doing in light of insight into their general tendencies (Ingram 2008; Krzywoszynska 2019; Morris 2006; Oreszczyn et al. 2010).

As noted in Chapter 1, different types of knowledge have been associated with different spatial scales. Tacit or sticky knowledge develops through doing alongside someone who has the expertise, suggesting that learning hands-on farming requires proximity to the expert (Carolan 2011). But in practice, growers do not always learn near another person and do gain expertise across distance, particularly as digital communication increasingly contributes to skills development. Digital learning in agriculture increasingly uses online platforms and social media to share information and peer learning, locally and globally (Klerkx et al. 2019). Growers use WhatsApp groups to exchange pictures of unidentified pests or ask for tips; on more formalised online networks, peers exchange advice or consult materials and advice. I was intrigued by these knowledge flows which seem to counter the received wisdom that co-presence is integral to agricultural learning because of the importance of learning tacit knowledge experientially. Practitioners and academics question whether digitisation can support knowledge exchange for all types of farming, including agro-ecological approaches (Klerkx et al. 2019). Carolan suggests that online exchanges are networks rather than communities of practice, as they cannot provide the embodied feel for how to do the skills of growing (2011). He is careful to avoid overstating Small Ag's reliance on tacit or local knowledge and recognises it differs by degree not kind from standardised or codified forms. But I think there is room to refine his interpretation of learning to grow, integrating networked communication and tactile practice just as growers do.

One grower described what he saw as the typical merging of digital information and hands-on practice:

> So my girlfriend, she's a seed grower and I quite often find her peering at her phone, looking at a demonstration video of, I don't know, how to put in anchors for greenhouses or you know, how to apply predators to plants or something like that. So, I think it is absolutely true that you will not be able to run a successful business by watching YouTube videos, but I do think that the fact that technology now is very transportable – and I'm sure she is not the only one who stands there in the field watching instructional videos while she's doing whatever it is that she's getting information on. So it's much more instant, much more accessible.
>
> (S3)

In essence, this is the same show-how-know-how model used in on-site induction, as the learner watches an expert and then mimics, repeats and corrects. Even face-to-face learning by showing involves a disjuncture as the watcher seeks to embody what they have seen. The expert's body cannot directly transfer their knowledge to the novice; the learner still must engage processes of translation as they work out how to replicate what they have seen, how to incorporate the movements they saw someone else make into their own movements. Dancers sometimes describe this as getting the choreography "into their body", a transfer which involves trial and error, and repeated practice. Emphasising learning alongside the expert and privileging being in the same place has perhaps implied that the knowledge transfer is more continuous than this, neglecting the significance of the stage during which the novice works it into their body, working it out for themself. Yes, it may be quicker to be corrected if there is someone who can watch and point out mistakes, but learning from YouTube indicates that proximity is not the key determinant of learning tacit knowledge; showing helps but *doing* matters. Asking one farmer about online videos as how-to guides, I asked about the notion that growing is best learnt alongside an instructor:

> I think face-to-face is important. I think you do have to get dirty as well. I think you actually have to have the contact with soil. I don't think you can do this purely from web analysis and go 'oh, now I know', because that just – when you're watching television, you're not getting wet or blown and you realise that you're not compacting the soil when you're stood on it and all of those [things].
>
> (S1.1)

His explanation suggests it is not contact with the instructor which matters but contact with the soil. The teacher might be virtual or physical presence – the real learning comes through doing the growing yourself, getting wet and dirty, realising what you've done, then doing it again and again.

Learning to adapt

One reason growers regard learning by doing as so important is that they know each farm, even each field, is different; no generic rules guarantee a good crop; their skill is fine-tuning practice to their specific context. Recall from Chapter 5 that the organic farmer had to learn how to transfer what he knew about growing on "boy's dirt" to his farm's "manly dirt", adapting principles to specific circumstances. Site-specific knowledge, typically gained through experience, is contrasted with more abstract or theoretical knowledge. Again, the classic Western hierarchy of knowledge is flipped, with that learnt through doing rather than in classrooms held in esteem, even by those with considerable academic training. A post-graduate degree in biological sciences, followed by university research in the same field, brought one graduate trainee into a large grower. His role draws on skills in scientific analysis and lab work to develop crop trials, but his scientific background does not equate to expertise in growing crops:

> I probably don't have the same innate ability to look at a plant and understand whether it's stressed or whether it's, you know, healthy, based just on visual experience that you would do if you were from a farming background, so to speak. But I think I can look at the data that's provided when you send off samples and look at the interaction between levels and understand things in that way, because I have a very quantitatively driven mindset. And so I suppose there's probably different ways to approach getting to a similar result. I think I can look at a crop and using, you know, an understanding of what has happened to that crop in terms of the environmental factors, you can understand why it may be healthy or unhealthy. But yes, I'd always want to be able to be – be more attuned to what's happening, you know, in the plant and be able to understand it better.
>
> (L2.5)

He contrasted theoretical insight into plant behaviour learned through lectures and in the lab with experiential knowledge gained through observing and working with crops, and innate skills acquired from growing up on a farm. Modesty may have led him to denigrate his abilities, but how he described his learning during his time in the company suggests otherwise:

> I think experience, to a degree of being out in the field a lot. But also, inquisitiveness and just having conversations with growers, asking what's going on, trying to understand what the operations are and what's happening, so that we can try and identify areas where we can improve. So, I think like, trying to understand why, for example, the fertiliser sprays go on at X, Z and Y points throughout the crop, and then trying to match that up to what we're actually seeing in the plant, and seeing in the data and the nutritional analyses and the plant development and all the rest of it, and understanding actually, is this necessary here, or can we reduce this by 30% here or, because the data suggests that that would produce a healthier plant and so – so I think a combination of experience and then engagement with, with the growers.

His account indicates that all knowledge types are valued and combined to develop optimal practice – theoretical and experiential insights work together.

Growers do not always implement generic practices informed by knowledge of plant needs because they are also looking outward from the plant. The strawberry farm's head of production does not necessarily implement the agronomist's expert advice because:

> They're very much looking at the perfect world scenario and what that plant needs in a perfect world. Whereas we're probably looking at it as more of an economical point of view, of yes, it probably does need that, but it's gonna cost me this much and we can still achieve a fairly good yield or quality without doing that, or doing it to a lesser extent. So it's sort of working in that factor as well.
>
> (L3.2)

Production decisions are not just about what plants need, but socio-economic variables too, which make mistakes a costly and informative learning process:

> The best thing in farming and being self-employed, you learn – the quickest way you learn is when you spend your own money and you make your own mistakes and then you don't make that mistake twice. And when you work for somebody else and you play with somebody else's money then, you know, it doesn't hit you so hard. But when you feel financial hardship because you've messed up, you make sure you're not gonna do it again, aren't you?
>
> (Su.1)

Part of learning by doing is making mistakes and learning from them: "I've been doing vertical farming now for seven years and all of my knowledge and experience is all from trial and error" (CEA1.3). Mistakes are inevitable when working with plants' complex variables.

Trial and error is a significant part of experiential learning, drawing on and developing growers' ability to adapt. But those who are largely self-taught are conscious this may be inefficient:

> one of the real concerns about learning here is that you sometimes feel that you're re-inventing the wheel and that's rubbish and I think that all of us who have started our own small farms and are trying to do this have to some extent re-invent the wheel or have had that conversation ourselves and realised we're having to do that and that's dumb.
>
> (S1.1)

This farmer would love to have a more experienced colleague to learn from, to challenge him to reassess practices, or bring inspiration to innovate. Autodidacts also lack transferable knowledge because learning focuses on particular sites and crops:

> I know an awful lot about the various pests and diseases that can attack cucumbers but my knowledge of potatoes is pretty minimal. So I've got a lot of knowledge but it is a little bit patchy and specific sometimes but it works for what we need.
>
> (L1.3)

This crop manager's significant experience is not accompanied by any qualification or formal evidence of his expertise. Several growers suggested the industry attracts people not suited to classroom learning, who seem to thrive with practical tasks and experiential learning. For the self-determined and independent, this works, providing they can access places to practise growing.

Having done his own internship, run his own enterprise and helped set up training programmes in agro-ecological horticulture, this grower offered a perspective on people following this route:

> There's a very holistic approach to why people get into growing on this scale. So, I was thinking that maybe that feeds into the kind of training and learning routes that people actually want, like, is there this, you know, not rebellion, but do they not want to sit in a classroom, do they not want to do a formal year-long training programme or go back to college and learn? That they've chosen this route, because of that, that kind of holistic lifestyle that they kind of learn as they go along and don't actually realise that there's an awful lot of formal training that you need to be able to do it well? And it's not 'til maybe a bit later, further down the line that they think 'actually, I need a bit more knowledge than

> looking at YouTube videos and like, articles online, and speaking to local growers'. And then you know, that's where maybe there's this gap that we need this more formal training.
>
> (S5)

The values which drive agro-ecological growers suit informal learning, through peer exchanges, and hands-on practice, which can leave gaps in their knowledge. But they persist, perhaps because it takes a particular kind of person to do this work:

> I was just thinking, you know, about the nature of what attracts you into growing and being a farmer or grower in the first place, often makes you want to do things your own way. So it's kind of almost inherent in a way [...] you know, growers are very pragmatic, so if they see a need, then they'll try and get it done as well.
>
> (S4)

They are the kind of people to find pathways into a job where few formal routes exist, to become auto-didacts, and to develop ways to train new growers; determination, pragmatism and adaptability are important skills in agro-ecological horticulture. Those in this community who believe that certain types of people find their way into the work perceive the self-taught, owner-grower of small agro-ecological horticultural as born not made. Like seasonal field-workers described as simply having "a knack", these growers are regarded as individuals who happen to have the right qualities for horticulture, essentialising and individualising their skills. But they do not develop their skills alone.

Learning by doing with others

The accounts of how growers learn shared so far highlight the significance of peer exchange, with doing together central to horticultural skills development. Flows of knowledge between workers within and between organisations are hugely significant to how growers know what to do. Peer learning is a widespread mode of knowledge exchange within agriculture (Sutherland and Marchand 2021). But talking to people familiar with horticulture, a common assumption was that people in the sector are not good at cooperating, that they tend to develop and protect their expertise rather than learning collaboratively. This could be one reason large growers voted to end the levy supporting AHDB – the last significant publicly supported knowledge institution for the sector. The structure of the horticultural industry works against centralised knowledge development and exchange because many businesses are so specialised and consolidated that they don't benefit from collective learning (Menary et al. 2019). A large specialist company develops expertise in its crops and invests in advancing its knowledge because it can, and because almost no one else in the

UK needs similar expertise. Generic advice services do not hold the insights they need; instead, they turn to academic researchers for innovations, either through publications or "we just go and literally pick somebody and ask them to come" (L4.1). This includes agronomists specialist in particular crops, and "internationally acclaimed experts in the field" (L2.5) paid to share expertise to keep growers up to date with the best production methods. Specialist growers are less likely to engage in peer learning because they have so few expert peers.

There are less transactional, more collectivised knowledge exchanges, via mentoring schemes or subsidised programmes such as Tyfu Cymru run by Lantra in Wales. This offered knowledge exchange visits to growers, and facilitated peer learning networks centred on crop specialisms, alongside training from technical specialists and experienced growers. Appetite for these activities showed that growers will collaborate, but there are divergent views about how much they are willing to share:

> One thing I've found in my time is how growers don't like to tell each other what they're doing.
>
> (L3.1)

This person's company does engage with other strawberry growers as the director speaks to them about marketing terms and cropping, and the production manager visits other growers. I asked how these exchanges avoid aiding competitors:

> We all wanna sell our fruit so we all sort of should be working together. But there are certain companies that just won't let you on their place and it's like, this is just ridiculous, what are you doing that's so different to anyone else? Nothing. So we're very open, like we will let anyone come in and we, we [are] very much encouraged to go and visit other sites and see what they're doing, literally from everything, from their structures to what plants they're growing, how they're growing it, to the irrigation strategy. Well when I go visiting I look at everything and try and gain as much knowledge as possible.
>
> (L3.2)

It is perhaps inevitable that those willing to engage with me represent companies more open to learning with others, less concerned they had secrets to keep. But commercial sensitivity is an issue, even for pro-sharing growers:

> We have a policy on our farm here that we are very open about the farming. So, you know, in normal circumstances, we would have several visits a year, people coming to us, and we will tell them everything really, you know, whatever they need to know; we don't try and hide anything. The only thing I would say we are a little bit cagey about is some of the

> business stuff. I think if it's any kind of cutthroat businessy type stuff, a little bit cagier, but the farming stuff, I mean, you're generally speaking, you know, you get back from what you – when showing a grower around your own holding, just from talking to them, you learn stuff.
>
> (L1.1)

Learning goes both ways, but there are limits to how much growers are willing to share with potential competitors.

Knowledge flows between larger growers may be impeded by commercial sensitivity and lack of peers. In contrast, agro-ecological growers have active learning networks closely associated with work to build a social movement (Anderson et al. 2019). The community of agro-ecological growers in many regards functions as a community of practice, as experienced growers train and support novices (Maye 2018). This apprentice style learning is crucial for the early stages of a grower's career and continues as more experienced growers learn from each other through formalised and informal learning. Membership bodies like the Organic Growers Alliance and Landworkers Alliance run events fostering knowledge exchanges and provide training tailored to members' needs. Visiting other producers is a common activity:

> Site visits are like probably the most useful. You know, that's like, I've found over the years that like probably the most useful information I've gathered, is from going and looking around other farms and like chatting to other farmers, and particularly people who you do get to know and sort of go to visit them once a year and kind of chat to them and reflect on what did work, didn't work. I found that really helpful over the years to have like two or three farms that I've gone back to regularly and had a kind of a sort of informal relationship with them.
>
> (S2)

As well as insights gathered on-site, visits provide growers with networks, enabling peer connections to exchange advice and ideas. Other networks arise organically, perhaps on a regional basis:

> I was starting out, and I was like 'I want to pick everyone's brains. We should have a growers group like Thames Organic Growers'.
>
> (S4)

In this case, a speculative email grew by word of mouth into a regional network of 100+ growers, sharing advice and arranging bulk buying. Online networks are popular, with specialist forums proving highly effective:

> I feel like my main source of knowledge now is I just go on UK Market Gardeners, ask a question and you get ten really good comprehensive

> [answers]. It's amazing the knowledge that's shared through that. So, yeah that's probably the most useful thing these days.
>
> (S2)

This grower described taking a photo of a problem with aphids he had not seen before, posting it to the group, then during a tea break looking over the responses on his phone, which included tips from some of the country's most established growers. Digital exchanges are highly integrated to how growers learn from each other.

Networks of growers represent considerable knowledge exchange, flows operating by goodwill – a form of gift exchange which works because so many growers see it as an important use of their time. Exchanging knowledge and supporting others is viewed as a defining characteristic of agro-ecology: "So pretty much everybody is desperate to share in their own way" (S4). This was associated with the character of people attracted to ecological food production:

> because there is a sort a kind of a comradery amongst growers, and you know, we all know how difficult a job it is. And when I used to go to the Organic Conference in the early years, you know, I was amazed at what an amazing set of people it was really.
>
> (L1.1)

Other growers referred to comrades toiling alongside each other with a common purpose and shared struggles. They also pointed to it being values-driven work, motivated by wanting to grow good, sustainable food. For some, supporting other growers is a political act, spreading the potential to transform food systems (Anderson et al. 2019). Supporting other agro-ecological growers does not create competitors but comrades:

> To me, if you were to ask me what my vision of a sustainable future is, [it] is that we don't need to become bigger and bigger. That's actually where everything's going wrong. What we need is lots and lots and lots more smaller people. And that's why, you know, I don't mind sharing information, because I want to encourage neighbours.
>
> (S1.1)

Exchanging knowledge and supporting new growers is part of the mission of agro-ecological farming, in a reciprocal sharing economy where those who benefitted from earlier generations pass knowledge on.

Informal knowledge sharing fills gaps in formal training provision and provides growers with the specialist learning they want:

> I think growers recognise that actually you only really learn about growing from doing it or from talking to other growers.
>
> (S2)

Working through a series of short-term jobs, gaining experience and learning from others, is a fairly typical pathway within organic farming in the UK and elsewhere (Claeys and Van Dyck 2022; Ekers 2019; Weiler et al. 2016). Many begin by volunteering through schemes such as WWOOF which places people on organic farms, where they receive food and accommodation in exchange for work. The use of volunteer labour is controversial and open to exploitation (Claeys and Van Dyck 2022; Ekers et al. 2016; Guthman 2014). Whilst not used by all organic farmers, it can help balance workload and costs. The crucial caveat is ensuring volunteers are learning whilst on a farm: "it's a knowledge exchange" (S1.2) in which the worker is "getting something back". When delivered well, the volunteer receives experience and knowledge in return for labour, as this former WWOOFer who now has a paid position as an assistant grower experienced:

> I find it really inspiring to go and like – it's just a really amazing way to travel, I think. I was kind of just not really sure what I wanted to do and was finding it quite hard to apply for jobs and have a proper job. And I wanted to learn. I did want to specifically learn more about food and how to grow food. And it's just a really good way of doing that for very little money really.
>
> (S1.3)

Volunteering has been useful for gaining skills growers require because the knowledge infrastructure limits other pathways.

Despite having an extensive history as a pathway into agro-ecological growing, learning through placements – voluntary, internships or paid – presents risks. Organisations which arrange placements on farms specify that they are cultural exchanges or "active vacations" not work (WWOOF n.d.). Helping hosts with a farm business is only permissible in return for mimimum wage (Workaway n.d.). Farms that take on volunteers or trainees should work within legal parameters for such roles or be liable for paying minimum wage. But as the term volunteer is not legally defined, there is a lack of clarity regarding its application (NCVO 2022). Some farms have treated trainees living within the household as family members exempt from minimum wage, but this exemption is due to cease (LWA 2024). Volunteers and trainees have reported conditions which are unfair, if not illegal, and can struggle to assert their rights to decent living and working conditions. In terms of educational value, there are risks with on-farm learning opportunities which are not quality controlled and depend on personal relationships:

> I think you're kind of lucky or unlucky on who you fall into train with, because when I think about how I trained, like all my basic knowledge is from that one person, X, and that still completely influences the way I think about growing. And I think I was lucky because actually I really like his way of working.
>
> (S2)

These opportunities are also inaccessible to those who cannot relocate, or live on a very small or no income, often excluding people of colour (Calliste et al. 2021). Those who gained experience through unwaged placements recognise this problem:

> I personally trained on just having accommodation and food and nothing, and that's kind of just how everyone seemed to do it when I was training. You just found a place and you volunteered for them until you knew what you were doing. But then other farms were very much of the opinion that people shouldn't have to do that, that we were privileged in a way, I guess, that we'd come from families that had money that could afford to basically go and work for nothing for people and they knew at the end of it, they could always go back and live at home until they worked out what they were doing.
>
> (S2)

This grower's CSA trains several volunteers each season, providing accommodation and a stipend to cover expenses. Some need to work alongside a traineeship to support families or other commitments. The CSA also faces financial challenges:

> I think actually, the biggest impediment for us being able to share our knowledge with trainees is actually finding the funding, so they can actually have an income while they're training. Every year, we're like scrabbling for bits of money to support trainees, we haven't got the money within our business to employ trainees.
>
> (S2)

Various grants and government schemes have supported traineeships but these are not guaranteed and sometimes carry restrictions such as age limits. These financial conditions reduce access to learning, another impediment to knowledge flows, exacerbated by the struggle to pay agro-ecological workers well.

Beyond traineeships' financial challenges, there are questions about unwaged farmwork within the organic community:

> over the years we've just had lots of discussions about the ethics of that, and for a lot of businesses it is essential, because they can't run, or they feel they're not financially viable without using volunteers, or they just want the community engagement or whatever else. But it's just a really thorny issue for – should the organic vegetable market be propped up by volunteer labour? Is it because we're not being subsidised elsewhere, you know, from Defra or CAP or anything else?
>
> (S4)

A small survey of OGA members found mixed views on the topic and varied practices around working with volunteers (Collyns 2022). Perhaps most revealing was that a significant number identified themselves as the ones not receiving minimum wage; during the phase of establishing the business, or even beyond, finances limit how much owners reward their own time and when this is divided across their considerable working hours, it can equate as little as £2 per hour. Economic pressures make it very difficult for agro-ecological growers to support fair work conditions (Dumont and Baret 2017).

These harsh realities of agro-ecological enterprise should, according to seasoned growers, be taught to new growers. The idea of an outdoor life growing good food appeals to many, but:

> that lifestyle isn't what it looks like from the outside. That it is 14, 16 hours, sometimes in the summer, it's hard work, it's cold and wet in the winter but actually, some people come into it a bit, you know, not really understanding.
>
> (S6)

To fully understand the realities of running a farm requires year-round training, to see the full seasonal cycle and understand all operations, and how tough things can be. This grower is sure to give trainees time working on data entry and accounts, to convey the harsh realities of running the business:

> I also think that certainly we're training new entrant farmers, there is no harm in them seeing that my wife and I are working seven days a week, 18 hours a day, because that's going to be their lot and they may as well because if they're serious about going into this sort of growing environment and this sort of thing [...] They're probably looking at the same things as we are, which is borrowing a lot of money, having the pressures of a lot of money – having families, still trying to balance that. I'm quite happy to share that. I don't think it puts a lot of the people off when they come here actually. I think it's also important to be realistic about what you can do. I think those are the things which are quite hard to teach but the pressures are there. It's very, very sad to see farms failing and it's no good if we started generation of new growers out there and don't go 'and it's really hard. It's not that easy'.
>
> (S1.1)

Hands-on experience of growing crops is part of growers' learning by doing, but perhaps it is more important to see first-hand realities of the lifestyle – long hours, low pay, high debt – to learn that it is not at all idyllic.

A significant drawback of the agro-ecological knowledge sharing economy arises from this: as knowledge is freely given, experts receive many demands on their time, typically receiving no financial reward for educating others. Exchanges are reciprocal or cyclical – learners become teachers – so knowledge is freely given, but some suggest expertise should be paid for. Time devoted to teaching is time taken away from growing, which has its costs. Many growers are prepared to bear this burden, but relying on their goodwill is risky:

> I don't know how sustainable it is in terms of, you know, burnout, and people come and go, you know, turnover of labour and everything else.
>
> (S4)

Very experienced growers offer an important resource feeding the sharing network:

> the really kind of in-depth knowledge about knowing your soil, knowing your land, comes from the same group of growers that have been doing this for a really, really long time. And if they no longer decided to be part of that group, if Facebook went down or what have you, then that knowledge would be gone.
>
> (S5)

If all those growers burn out or go out of business, it is not clear how their knowledge could survive. Peer exchange so crucial to agro-ecological growing depends on the expertise and goodwill of a network of experienced growers, which in turn depends on them maintaining viable enterprises, neither of which conditions are assured, given the pressures on them and their businesses.

Conclusion: The capability to become a good grower

What you need to know to produce plant crops is a combination of plant-centred techniques of tending, and more outward looking abilities to understand the operational context. A skilled grower weighs up these complex variables and judges which actions will produce a quality, profitable crop. Whilst many aspects of this know-how have parallels in other industries and with other forms of practical or manual work, there are specificities associated with the biophysical properties of plant products; seasonal growth cycles and perishability make timeliness and timing particularly important. These interact with movement of labour and costs to present pinch points which are made particularly tight by the low prices paid to producers for most fruits and vegetables.

Despite the complexity and knowledge intensity of growing good crops, the skill of this work is underestimated even by those who work in the sector. Although many do describe it as skilled work and appreciate their workers' abilities, including "unskilled" fieldworkers, the nature of the tasks tends to narratives which understand it as a bodily knack rather than a learned ability. It is too easy to assume that good growers are born not made, obscuring the learning it takes to become a grower, imagining horticultural skills as facets of individual bodies. These misconceptions are also fed by the sector's reliance on learning on the job, which is undervalued as a source of specialist expertise. Many good growers, including technical experts, developed their skills through doing the job, with learning through showing and doing key flows of horticultural knowledge. The centrality of experiential learning and value placed on know-how gained through hands-on experience tend to support the assumption that horticultural work requires practical ability more than intellectual knowledge. Such bodily intelligence is easily overlooked and regarded as lesser than intellectual knowledge, contributing to social undervaluing of horticultural work. Attention to how growers learn shows that all know-how blurs distinctions between practical and intellectual abilities, so job categorisations reliant on them should be reconsidered.

Far from being individuals with some innate knack, good growers have taken time and effort to learn wide-ranging skills, developing specialist know-how rooted in experience. The prime mechanism for learning how to be a good grower is "show-how" – demonstration by an expert, mimicked by the novice who then repeats, and repeats, improving and learning by doing. This learning suits the hands-on, place-specific know-how growers require. It also indicates that horticultural know-how is thoroughly social, developing through engagements with others, existing as practices done together, either side by side in fields or across shared digital spaces. Although the stories people tell about learning to grow suggest horticultural aptitudes rest within an individual, they are collective and learnt: good growers are not born but made together.

For large-scale conventional growers, the benefits of knowledge exchange are constrained by specialisation and competition, making them less likely to learn from each other. Other types and scale of growers engage in learning exchanges as values-driven contributions to the movement to produce good food. The extent of their knowledge sharing also emerged as a pragmatic response to the lack of formal learning opportunities in agro-ecological growing. But access to these flows is not evenly distributed, with some would-be growers struggling to access opportunities such as traineeships. The capability to develop horticultural know-how is also dependent on access to places to practise growing, which takes land, capital and agency. The power to grow food is a crucial element of the conditions which enable horticultural knowledge to flourish because the know-how to grow food only exists through the social practices of working to grow food. And these powers extend way beyond the inward world of the plant, so can be highly constrained by wider circumstances. The knowledge infrastructure growers draw on and how they are limited, particularly by economic constraints, are the focus of the next chapter.

Note

1 Fertiliser Advisers Certification and Training Scheme – a validated programme to regulate fertiliser use.

7 The health of horticulture's skills ecosystems

Introducing skills ecosystems

So far I have shown that skill is a socially embedded practice, combining thinking and doing, located within and between bodies, across environments which include organisational systems, materials and technologies. The previous two chapters featuring practices across my encounters with growers illustrate the wide-ranging know-how comprising horticultural skills, with good growers being able to grow food, run a viable enterprise and adapt to change. They look inwards to the doings of plants and outwards to the world consuming their produce. The complexity of the work and skills development are too often overlooked or misunderstood, rendering many horticultural workers "unskilled", perhaps with a "knack" for their tasks, that is born not learnt. Becoming a good grower entails learning by showing and doing, exchanging knowledge with peers, colleagues and experts. But things get in the way: knowledge flows might be impeded by poor training infrastructure or lack of peers with the right know-how. In this chapter, attention turns to the system of knowledge flows around a grower, situating them amongst broader contexts. This helps understand the nature of what the sector has called its skills crisis, attending to constraints on growers' ability to become skilled.

By attending to the flows of knowledge into and around horticultural production, it becomes clear that a skilled grower depends on other people and materials; their learning sits within and draws on networks of things and activity. But various blockages impede knowledge flows, and sources are lacking or depleted. Blockages, leaks and barriers affecting knowledge flows featured across the description of growers' learning, as did limits on the capability to develop horticultural skills. I discussed these flows and impediments with growers and others, seeking their perspective on knowledge networks; during these conversations, I began calling them horticulture's knowledge ecosystems, conveying systems of actors, resources, relations and settings. This terminology seemed to make sense to people used to working in ecological contexts where habitats and species interact, exchanging resources influencing each other. So it became a useful metaphor to explore access to the skills required to grow food; later I discovered that scholars have used similar metaphors to

DOI: 10.4324/9781003281689-7

conceptualise skill and education. Finegold coined the term skill ecosystem to describe geographically or sectorally clustered organisations needing staff with advanced skills (1999). He drew parallels with ecological systems for their dynamism and relational nature, mirroring complex interactions within skills development (Brown 2022). Later expanded to cover other types of work, skills ecosystems were proposed to overcome tendencies to analyse labour markets through focusing on either supply or demand, by considering how the two sides are connected (Dalizel 2017).

As defined in the literature, a skill ecosystem comprises business settings, policy frameworks, modes of engaging labour, job structures and systems for skills formation (Buchanan et al. 2017). The metaphor sets skills and education within their broader socio-economic context (Dalizel 2017). This suits an ecological definition of skill as it supports multi-scalar analysis considering the micro detail of the skilled individual within their meso and macro context. To understand skills ecosystems means considering what is constraining and enabling skills development, addressing the effects of power and inequality (Brown 2022; Dalizel 2017). Another appeal of skills ecosystems analysis is its recognition that profit and skill are mutually reinforcing: to be productive, workers need skills, and to offer skills development opportunities, employers must be profitable (Dalizel 2017). In horticultural businesses, the capacity to develop workers' and trainees' skills is highly dependent on financial viability, with the profit squeeze common across the sector constraining growers' ability to invest in knowledge flows.

Although coming from a different academic tradition, skills ecosystems also suggest a systems perspective which considers complex, multi-scalar interactions between ecological and social systems, recognising non-linear complexity, features and feedback (Folke et al. 2010). A systems approach regards food as the outcome of complex socio-ecological processes which interact in multiple unpredictable ways, resulting in complex problems (Ericksen 2008; Pereira et al. 2020). So a systems perspective on horticultural knowledge considers the complexity of socio-ecological processes and interactions at play. As indicated in earlier chapters, problems associated with horticultural skills arise from flaws in broader agri-food systems; agri-food challenges are complex and systemic, so their skills dimensions are too. Making sense of systemic challenges means considering actors and processes, and interactions between them, without expecting simple cause-effect dynamics. A skills ecosystem perspective suggests failings arise from processes becoming misaligned, or uncoordinated skills development practices (Buchanan et al. 2017). By following flows of horticultural knowledge and how they become blocked or misdirected, it might be possible to understand the sector's problems (Pitt 2021b).

My analysis of horticulture's skills ecosystems is informed by this academic literature, whilst regarding the term as heuristically helpful metaphor more than absolute concept. In line with the ecological definition of skill, the terms knowledge and skills ecosystems are interchangeable. I propose a skills or knowledge ecosystem comprises actors and flows relevant to using and developing knowledge, highlighting how knowledge flows are impeded and enabled. One goal of characterising horticultural skills ecosystems is to understand how healthy they are, and where they might benefit from restoration, addressing the sector's skills crisis. Like the biological counterparts the metaphor references, healthy skills ecosystems have resources – capital, a supply of trained workers – which enable businesses to grow and develop (Brown 2022). Actors should live in an environment that supports skills, for example training infrastructure and interdependence (ibid.) A flourishing ecosystem is diverse and healthy enough to support the life dependent on it, meaning a healthy horticultural skills ecosystem services the knowledge essential to abundant food growing. Knowledge flows are unimpeded, all actors have adequate resources to feed knowledge exchanges and access to the system is open to all. And like any thriving ecosystem, it can adapt to change – it is resilient. Social and ecological systems are dynamic, and change is not always predictable, so they need to respond to change and evolve into new forms (Folke et al. 2010). A resilient skills ecosystem can self-correct following shock (Brown 2022), and take new forms as required, so we need an eye on the future.

Mapping skills ecosystems

There is no precise methodology for mapping skills ecosystems, rather it is a heuristic process uncovering resources and flows which production relies on, within the wider landscape. I analysed data from numerous growers to devise archetypal knowledge ecosystems, starting from a grower or farm and working out to knowledgeable actors they interact with and the context they learn within – their habitat. The ecosystems presented here connect to my definition of horticultural skills and its three categories of knowledge: growing food, managing the enterprise and R&D. By becoming familiar with growers' operations and discussing them with numerous actors within and beyond their production system, I identified key "knowers" informing growing and characterised how knowledge flows between them. I discussed initial versions of the ecosystems with stakeholders and then refined them. Since the ecosystems were devised, the context has changed, for example the dissolution of AHDB Horticulture in 2022, but they feature on the maps as growers at the time viewed them as important.

Two skills ecosystems are presented to highlight key contrasts across the sector, those of larger and smaller growers. Having considered what distinguishes growers involved in the research, and how this affects knowledge processes, it seemed the most significant line of difference was that of scale. The operations of large scale and small or family scale organic growers are quite different, although they share many values and techniques. Conversely, conventional and organic growers at scale use diverging production techniques – although tightening controls on pesticides and herbicides push them closer together – but logistically are quite similar, particularly regarding accessing expertise and workforce development. The characterisation of larger and smaller growers is imperfect and imprecise, but useful for exploring knowledge challenges. Significant divergences between growers at each scale or between production systems are noted. One question raised during stakeholder reflection on the draft ecosystems was whether there are any "medium" growers in the UK. Whilst there are some falling between my two archetypes, the country's horticultural industry is fairly bifurcated between those who have specialised, intensified and scaled up and those operating small, single sites supplying alternative supply chains (Chapter 3). A key scalar distinction in relation to knowledge ecosystems is workforce size, as the two sides of the bifurcation point to very different labour models: significant teams of seasonal workers in larger scale growing versus minimal waged workers in smaller scale growing. The largest number of employees at any grower I researched was in the thousands, compared to one part-time grower employed on a family farm.

The archetypal knowledge ecosystems are presented in turn, addressing: how does knowledge flow, who/what is involved and what impedes or enables the flows? Like any map, they are imperfect and only show a moment in time; the actors depicted could extend back through time and interactions to show who each actor learnt from, and who they learnt from. A pragmatic resolution was to start from a grower and focus on those feeding them knowledge. Before presenting each archetype, I take you into examples of growers they refer to, through a vignette highlighting knowledge practices particular to the scale. These bring to life portraits of horticultural skills sketched so far, whilst highlighting skills essential to their operations beyond those obviously associated with food growing. First, we visit a family-owned organic farm operating at market garden scale represented by the smaller grower skills ecosystem.

At the kitchen table: Feeding the global family

"Dinner's running late!" was the call that greeted me as I let myself in the front door just before 8:30 pm, the time I'd been told to come. Tom was still out in the fields sorting something, whilst Siobhan guided the volunteers in dinner preparations. They were joking and chatting, clearly getting on well despite being recent arrivals. My offers to help were rejected, so I perched on an armchair with one of the cats. Siobhan was muttering to herself as she made the pizzas, switching between tasks as something else came to mind. Tom appeared in his work clothes, minus boots which had been swapped for slippers. He clocked that there was not enough bread left for the next day, so reached for flour and scales. Phone rang, and from his end of the conversation, it was clear his eldest daughter needed picking up. And just like that I became a taxi driver.

It was well after nine when we sat to eat. The kitchen table had just been cleared of piles of paperwork and laid for eight: the family, residential volunteers and me. Pizzas were on the countertop so we could choose our toppings – veg from the farm, various cheeses, salami. Wine and salad were passed around, amidst reminders to help ourselves to seconds. As we ate, Tom and Siobhan told stories of raising baby plants in their bedroom. Then talk turned to politics, particularly around food production, everyone agreeing how screwed things are and the need for protest to bring change.

After everyone was well fed, Tom said several times that he needed to go to bed, but kept struggling to drag himself away. Finally, around 10:30 pm he and Siobhan headed upstairs, declaring the washing up tomorrow's problem. Less than five minutes later Siobhan appeared in her dressing gown, checking some detail, looking for a crucial piece of paper amongst the heaps cleared off the kitchen table. Then Tom reappeared, also amidst some administrative task to sort before sleep. By the time they properly retired, it was not much more than 6 hours until Tom would be up to start the next day's work.

Next morning, Tom and the volunteers were out in the fields when I returned to the kitchen to find breakfast. Siobhan and I chatted until break time when the volunteers appeared, put the kettle on and carved off thick slices of just-baked bread. Then an air of mild panic as Siobhan realised there was nothing for lunch. She began rifling through fridge and cupboards whilst throwing out ideas of how the children could be occupied during the imminent school holiday. Break over, I headed out with the volunteers to pick beans and carrots, under Tom's careful instruction that last week's bunches had been too generous – not too many big ones this time.

I always enjoyed staying at the farm, the food and conversation felt nourishing, the welcome warm, be it dinner around that big table or on lap-trays round the TV. Tom and Siobhan are great hosts – they have to be as they usually have a full house, hosting two residential volunteers across the growing season. Most come via WWOOF for between two weeks and several months, working in exchange for bed and board. The family see looking after the volunteers well and offering experience of farming as a fair exchange for the work. Part of this is sharing an insight into the food system, hence frequent dinner discussions of the wrongs of Big Ag and the rights of local food. They also love food, so want to eat well and feed others well. But this creates work, including emotional labour, as Tom described:

> A factor that doesn't maybe come into the conversation which should is the more people we have here, the more people who are living with us, so that's not only the number of people I'm cooking for, which I enjoy, I don't begrudge that, I like that. But you're cooking for maybe ten, 11 people lunch and dinner every day. What you also don't see on that is that everyone wants to talk to you about food and what we're doing all the time and that's part of the deal. And so it's fairly unrelenting and quite regularly – because we have a turnover of people – having the same conversations. I am still engaged with that conversation. I find it interesting and I like to engage people that maybe haven't had those conversations but it's nonetheless tiring for it.

Volunteers stay in the family home which adds pressure:

> It's the nuance of the relationship as well, which is complex because we also live together. So, those people are in your living, breathing space, so the strain on, you know, on just your internal relations as well, it sometime is – so people are very different within those groups and sometimes it's really easy and other times it's really quite tiresome.

They described themselves as sociable and are glad the children meet people from around the world whilst living in a relatively remote spot, but living with others is a strain. Most are lovely and do a good job, but "then there's a quarter who you just can't wait for them to go". Siobhan gave a memorable example of this minority, laughing as she mimicked this infuriating volunteer floating around the kitchen not even lifting a tea towel whilst everyone else mucked in:

> He stuck in my head because I had friends in one week, and so because I knew they were staying down in the cottage I baked basically two loaves of bread the night before and two loaves of bread the following morning, okay? And I came in for lunch at noon or whatever, one pm, and there was one loaf left. I was like, 'what's happened to the other three loaves?' And they were like, 'oh, we've eaten them'. I was like, 'three loaves? In the course of three hours?' I lost it with them. I was like 'eat something else! There are other things to eat than bread'. And apparently everybody 'fessed up and said it was largely him, you know, the young eighteen-year-old boy. You can't kind of blame him, yeah. So having volunteers can be testing sometimes.

She was glad the current volunteers were easy to get on with and motivated, one of them keen to gain experience on many farms. Tom's careful sifting for applicants really interested in sustainable food seemed to have worked.

I'd always sort of known that small farms entail hard work and long hours, a different kind of family life. But seeing it up close, negotiating piles of washing up to be able to rinse the salad for dinner, tallying how little sleep Tom had and talking to Siobahn brought home how much of the crucial work of running the farm happens not in the field, but at that kitchen/office table. That is where people are fed, the bookkeeping is entered on a spreadsheet, the Airbnb requests are replied to, where she does many things for the farm:

> That's the other struggle: I manage way too much. I largely manage the kids, I do a lot in the household as well as working two days a week, which gets spread over the entire week, for the charity, and then trying to stay on top of our books and our bookings. It's too much to manage. Even with a bookkeeper, and we do have accountants but we kind of keep them to a minimum because it's expensive.

She rarely helps in the fields, although will pull on wellies when necessary, but without her work the farm could not function. She lacks Tom's knowledge of how to grow a great crop of carrots, but he is not very numerate so needs her computer and financial skills, plus the diversified income she brings in. Workload creates pressure; then there's the loan used to buy the farm. When I asked what their biggest challenge is, her answer brought all this together:

> The financial stress of it, the stress of running around, Tom's stress. I mean, but that's also been a feature of my life in general. I think it's a feature of most people's modern life and it's just really wearing you know? So I would say that. That and then staying on top of the books, you know? I constantly have, you know, I need to deal with things and yeah – but that's part and parcel of the whole … you know what makes it stressful is that there's never enough time, you know?

Looking for some hope, I asked if there was a prospect of taking on a grower to share the workload:

> I like to think that in the future we'll grow the farm to a point where we might have one, you know, and that's part of what that expansion is. I mean, I think in an ideal world what we'd like to do is continue maybe growing on the level which we're growing but to also add in training and to start to make some income from that, that might allow us then to hire one or two people who might manage that. But whether it ever pans out that way, I just don't know.

Increasing farm income to cover more salary is tricky as it takes significant labour to generate more profit. So longer term?

> You know, and that's something we think about, and Tom thinks about a lot because he's tired now. He's really, deeply tired and we're both now in our mid-fifties. And we're both pretty tough, but, you know, but the exhaustion of it. I mean, it gets to him more than me, but I've found myself really tired in the last year. So I don't know, you know. We're stuck because we have to, I think we still have – our loan is another ten years we have to get through to, you know, to pay this off. And if we can do that then I think we want to see this stay a farm, you know, but we'll have to figure out how we're going to work with that as we get older. Because I don't know … I mean, maybe, I don't know what kind of world we'll be living in, you know, and maybe the children will be interested in taking on the farm but, you know, they also may well not be and then what does that mean? You know, yeah. But I mean that's something that all farmers face, you know?

> I think we both feel like we're doing something that's important. You know, it may be small and tiny and not that impactful on the big – but we feel like it's a really important thing to do. And that pulls you through all the stress and all the other, you know, stuff that goes with it.

So alongside all the know-how and sheer hours of toil it takes to grow vegetables, determination keeps the business going through the emotional and financial stresses.

The skills ecosystem of smaller growers

"Smaller growers" indicate operations owned and run by one or two individuals as a stand-alone concern, such as a market garden, or family farm like Tom and Siobhan's or a community enterprise such as a CSA. They are small in area, with many horticultural enterprises in the UK falling below the 5 ha eligibility for direct farm payments under the CAP. They are also small in workforce, typically with fewer than ten workers, and often none beyond the farm-owning family. Some describe this as "human scale" farming as sites and methods are designed around the people working them, not for large machinery. Whilst not all use volunteer labour, this is a feature of smaller scale growing in the UK, particularly for those operating to agro-ecological or organic principles. This is partly because working with these growers is a key educational pathway, but also reflects the difficulty of paying the workforce a living wage. Tom would love to pay everyone who works with him more, but can't afford to. His farm is fairly typical of this scale as it focuses on local or "alternative" supply chains through their own box scheme, on-farm retail, local independent markets, retailers and hospitality. These customers want a mix of produce, so the farm grows diverse crops, in-field and under-cover. Tom is responsible for everything from crop raising through to sales and marketing, and all associated processes.

Table 7.1 describes the key elements of a smaller grower's knowledge ecosystem – know-how, knowers and knowledge flows – reflecting what these growers need to know and how they access it. The key actors using and developing this know-how, and exchanges between them are presented in Figure 7.1. Together these suggest some distinguishing features of this knowledge ecosystem and its challenges. Firstly, the amount and range of expertise required is shared across a relatively small team. Operating with one or two workers means little division of labour, and as these growers are likely to produce a range of crops, each needs wide-ranging horticultural knowledge. Secondly, knowledge flowing in often comes via processes such as peer exchanges, for which no money changes hands. Growers are likely to have learnt to grow through training placements with more experienced growers, or while doing the job. Tom's assistant learnt at various community gardens and through time as a WWOOF volunteer. Such learning is supplemented by self-teaching and experimentation, and resources such as books

Table 7.1 Smaller Growers Skills Ecosystem

	Growing Food *Know-how to produce plants*	***Managing the Enterprise*** *Know-how to maintain site & operations*	***Research & Development*** *Know-how to enhance productivity and minimise risk*
Description	Small team responsible for producing diverse crops Knowledge specific to location and production system Emphasis on soil fertility Chemical-free pest and weed control	Whole supply chain management including sales, marketing, customer service Financial management, e.g. bookkeeping, grant applications Hosting residential volunteers and running the household	On-farm trial and error Trials run by specialist organisations and research institutions
Key Actors	Grower/Head Grower Employed or volunteer Assistant Grower Apprentices/trainees	Partner/family Bookkeeper/ accountant	Sector organisations, e.g. OGA, Soil Association More experienced growers
Knowledge Flows	Learning by doing Showing by more experienced growers Advice from peer networks through informal exchanges or mentoring Online and printed resources	Learning by doing Advice from peers through informal exchanges	Advice and training sessions, e.g. Farming Connect Publications, e.g. OGA magazine Online forums

and videos. Thirdly, organic and agro-ecological approaches common at this scale entail work to adapt general principles and practices to specific sites and contexts, making learning by doing or trial and error inevitable. Fourthly, many smaller growers actively participate in the community of practice represented by grower networks and organisations such as the Organic Growers Alliance or Landworkers Alliance, both of which Tom has been active in. In addition to coaching volunteers and his assistant, Tom is involved in initiatives to develop more formalised training, mentoring new entrants and hosting farm visits. A national training organisation recently invited him to deliver specialist short courses, for which he gets paid. Before this, all his expertise was shared voluntarily, because he was committed to furthering agro-ecological farming. A key feature of this ecosystem is how growers'

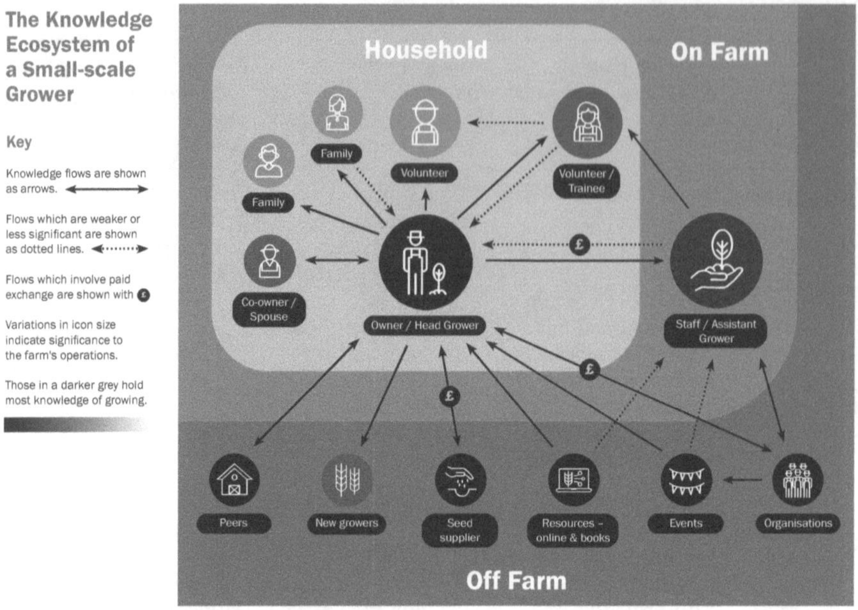

Figure 7.1 Smaller growers knowledge ecosystem.

social, environmental and political values compel them to grow agro-ecologically and to act as advocates or educators within the movement. Personal commitment to growing good food provides important motivation for hard work, sometimes for little or no wage; values-driven commitment to growing also leads people to seek entry into this ecosystem.

Table 7.2 highlights features impeding or enabling flows of knowledge to and between actors in the smaller grower's ecosystem. Some impediments arise from the nature of food growing (e.g. unpredictable weather) or specific challenges of agro-ecological production (e.g. volunteer turnover). Others arise from the nature of a small business with narrow profit margins (e.g. limited capital). One stakeholder suggested some features typical of any small owner-operated business. Workload and self-exploitation are not unique to horticultural enterprises and have long been features of family farms (Galt 2013; Gray 2014). But as indicated in Chapters 5 and 6, particular features of these operations make challenges more acute. For example, the importance of timeliness when working with fresh crops and the number of unpredictable variables when working with plants increase production risks. Next, I delve into the challenges smaller growers face, to see how healthy or resilient their skills ecosystems are.

As Siobhan's feelings suggested, one word summarises the pressures smaller growers face: time. There's never enough time. Growers described long hours with little time off, resulting in physical and emotional strain, which makes it harder to think through solutions. One grower at a small cooperative joked

Table 7.2 Knowledge flows in smaller growers ecosystem

	Growing Food *Know-how to produce plants*	***Managing the Enterprise*** *Know-how to maintain site & operations*	***Research & Development*** *Know-how to enhance productivity and minimise risk*
Knowledge Impediments	Uncertainty & uncontrollable variables, e.g. weather Adapting generic knowledge to a specific site Difficulty recruiting experienced growers Volunteer placements entail regular turnover	Lack of personal aptitude	Lack of time to engage Lack of resources to invest in innovation Unpredictable future Limited provision of relevant specialist knowledge exchange
	Workload exceeds reasonable capacity Lack of resources for learning Lack of time for communication Crucial knowledge held by one individual Limited record keeping/data Autodidact learning is inefficient Working under physical and mental strain		
Knowledge Enablers	Personal qualities and motivations, e.g. passion, resilience Willingness to share knowledge and support other growers Informal learning is viable with minimal financial resource		

that if the work of growing is fun and relaxing, you are a gardener; if it's stressful and demanding, then you're a farmer or grower. Having a small, or no team, places many responsibilities on a few shoulders, so when a tractor breaks:

> I feel an emotional strain, it's like I'm so busy already and suddenly the only person that's gonna fix this is me. And then I have to go and learn how to weld and there's that horrible realisation that ends up – I'm just gonna have to go and learn how to do this.
>
> (S1.1)

The benefits which hobby gardeners often enjoy (Pitt 2014) do not scale up when the operation grows, or when the crops need to generate income.

Even within small teams, growers noted that it can be difficult to keep information moving due to workload. Regular meetings and recording data or plans would be useful, but slip down their priorities; plus those attracted to growing do not always have an affinity with such tasks which exacerbates the risks of relying on key individuals:

> I think probably the majority of small farms run like that, that there's one or two people who hold all that information, and it can very quickly all fall apart.
>
> (S2)

This is risky as people may be incapacitated or leave:

> I guess the main vulnerability is just how it's proportionally much more dependent on individuals than in a larger business.
>
> (S4)

It also exacerbates the strain on key people: "it's a lot of work and a lot to keep in your head and a lot to keep on top of" (S2). The breadth of each worker's responsibilities makes it difficult to replace them and creates risks of burnout.

As well as the range of tasks growers bear responsibility for, growing with few mechanical and chemical inputs is labour intensive (Timmermann and Felix 2015). A globally influential figure in agro-ecological farming, Jean-Martin Fortier, calls it the replacement of mass farming with farming by the masses (2014). Tasks like weeding are crucial and almost constant on agro-ecological farms, requiring relatively high levels of physical labour done by relatively few people as growers seek to minimise wage costs. The price of property and land in the UK not only presents a huge barrier for new entrants, particularly those from marginalised backgrounds (LWA 2017); it also means many businesses are established through debt. The strain of too much work and too little time is made more acute by financial pressures:

> How do we make sustainable farming pay? Because, it's evidenced widely in the States – there was a huge number of young farmers who've gone into farming who can't make a living off it. And we have to fix that. But again, I think that means also changing our economic system so the likelihood of that happening…
>
> (S1.1)

The dominant agri-food regime keeps agro-ecological growers in niches where they struggle to remain profitable. Scaling up production might seem an obvious route to increase profitability, but:

> The problem is scaling up and the finances of scaling up. It's labour really over everything. It's just having the money to invest in that initial scaling up and then getting it delivered.
>
> (S1.1)

This farmer would love to take on more staff and pay them more, but growing the business to generate more profit generates more work which cannot easily be covered by the workforce their economic capacity can carry – a tricky balance between workload and income. Plus it has not always proved possible to recruit: "I've been desperately looking for an assistant grower, and the difficulty of finding that person, that has surprised me actually" (S1.1). Whilst there is a movement of people into this sector (Taherzadeh 2020), the pool of people skilled in commercial growing is limited and they struggle to find affordable accommodation in rural areas (Styles et al. 2022; Nye and Lobley 2021).

The balance between capacity and income is often made up by voluntary labour, although this is also driven by social and educational goals, particularly for community focused growers who want to engage people with food production. However, this is not an uncontroversial issue and is often debated amongst growers, as already highlighted. Minimum wage legislation means: "[if] all you're after is free labour, that's not only not particularly moral, it's also actually illegal" (S3). Long-term volunteering is therefore offered in exchange for learning and experience:

> we are aware that we definitely run on that volunteer help to make it viable. I think that is true. I used to feel really concerned about that and I feel less concerned because I feel that we are able to pay back in a different sort of way. And that's not hollow words. We really put a lot of effort into doing that and I think people appreciate that.
>
> (S1.1)

Feedback from their volunteers suggests this exchange is valued, but not all hosts provide such positive experiences as evidenced in online forums where travellers complain about poor treatment or accommodation and discuss whether WWOOF is exploitative.

Volunteering is a key pathway into agro-ecological growing, with many experienced growers having taken voluntary roles to learn the job, but this knowledge flow entails significant impediments. The opportunity to train is restricted to those able to take a low income: "if you can't pay people there's only very few people who can do the training" (S2). Aspiring growers may also need to relocate for a short time, excluding those with geographic ties such as caring responsibilities. Being able to secure external funding to pay trainees is another skill some smaller growers develop, and part of their workload, without which this skills ecosystem would be weaker.

Internships and training placements in agro-ecological growing are a self-organised corrective to the long-standing inadequacy of mainstream provision. But this system also arose because it is perceived as a good way to develop the skills growers require. Growers form communities of practice in which novices are inducted by the more experienced (Lave and Wenger 1991; Wenger 1998). Knowledge is exchanged within growing operations and between growers: "just sort of learning as you go along" (S1.1). Exchanges occur through networks of smaller growers, farm visits, events and discussions which are increasingly mediated by digital exchanges. Whilst there are advantages to learning by doing, it has limits: autodidacts may not do things the most efficient way, or might become expert more rapidly through training. Growers also lack time to innovate: "I don't even have the time to think it through properly, let alone make it happen" (S1.1). Working time is precious, so accessing knowledge exchange is difficult. Peer networks and informal knowledge exchange depend on people giving their time, which growers are always short of; the ecosystem is vulnerable to shortages of this essential resource.

The voluntary nature of many knowledge flows in this ecosystem is both strength and weakness. The strength is that flows operate independent of cash transactions, meaning they can operate without external investment. This is underpinned by growers' commitment to help others, with a system operating as mutual aid with those who benefitted from others' expertise wanting to offer the same. Know-how is shared or collective rather than individually owned (Anderson et al. 2019). But these facets also weaken the ecosystem. Experts' time is a precious resource, and exchanges depend on their capacity; knowledge freely given might be over-exploited, risking burnout. More formal programmes run by the Soil Association, Landworkers Alliance and others depend on project funding which is short-term and uncertain. Some argue reliance on the voluntary sector is wrong:

> Actually, I don't think it's a charity's job. I think it should be a government thing and it should be linked very closely to a comprehensive strategy to get a resilient, sustainable food and farming system.
>
> (S3)

Growers who have accessed government support to host paid apprentices suggested this as a permanent model offering a viable pathway into growing, increasing their capacity to invest in it fully and fairly. This requires either more profitable businesses which can pay for additional labour or external funding to remunerate the time they contribute to educational activity.

In the smaller growers' knowledge ecosystem, it is striking that potential areas of strength and vulnerability overlap (Figure 7.2). Voluntary skill sharing provides a specialist, hands-on training system, but limits entry to trainees who are mobile and can live on low incomes. Should this be a feature of just agroecological food production? A sharing culture of knowledge exchange fuelled by goodwill can function in the absence of external funding or institutional support, making it resilient to changes in policy or investment landscapes. But it is vulnerable at the level of the individual as the trainee is not receiving fair pay, and the trainer can only teach when they have time and energy, which is subject to their workload and economic viability. The strength of the ecosystem lies in the connections between actors, but the risks sit with them as individuals.

Some growers have structures to insulate against these risks, for example cooperative and community models which distribute workload, and harness customer inputs through financial or sweat equity. Research participants suggested such collaboration and collective advocacy for greater public support

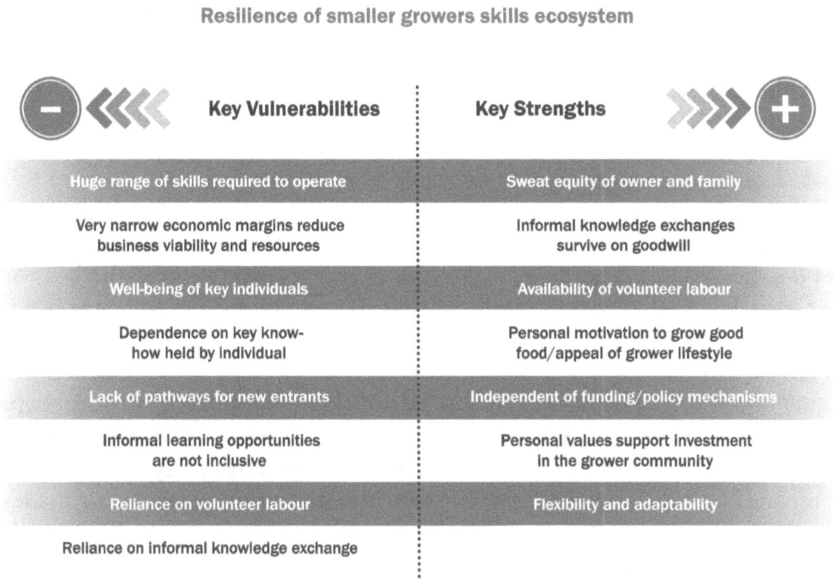

Figure 7.2 Resilience of smaller growers skills ecosystem.

are key to allowing smaller growers to thrive; they aspire to networks of connected growers sharing equipment, staff and expertise, combining crops to provide a large or diverse supply. Such division of labour would gain advantages of specialisation and volume which can come through scaling up, whilst retaining human-scale operations and values. Before reflecting on these matters of scale, detail on the larger growers skills ecosystem begins with a view from a second farm.

People skills: Getting the best from the team

"We're not a business that tells people to do things anymore" was how Connor put it. The company where he is farm manager and has worked for almost two decades prides itself on looking after its employees. It seems to be working as they have awards for being a good large employer based on international standards. Their HR team told me that "serial appliers" pursue pretty much any vacancy because they're so keen to work there. Like Tom and Siobhan's farm, theirs are certified organic, but the operations couldn't be more different. The company's farms supply a large organic box scheme, meaning a significant workforce spread across farms, polytunnels and offices, plus packing houses and drivers. The main office building had a nice feel, with fresh décor and baskets of fresh fruit; people were chatting and smiling, signs promoted upcoming learning activities and reminded everyone to vote for new employee representatives. Perhaps it's easier to look like you're enjoying your work in the shelter of a modern office building – the real dirty work is out on the farm. But whenever

I went to one of their fields or polytunnels, I was struck by the calm and quiet, the absence of a sense of urgency. Everyone across the farm team emphasised the importance of speed and efficiency and conveyed a constant sense of pressure to deliver, but this didn't manifest in shouting or hassling teams to go faster.

So if they "don't tell people what to do", how does the work get done fast? Connor explained that when he started in farm work, labourers were treated "a bit like a number", so they were told "you need to pick this and you need to pick it quickly", and not much more. But people are not like robots, he explained, so need a more nuanced approach to get the best out of them. The key is "people skills" – what all good managers and supervisors have. The qualities of people skills are difficult to pin down, perhaps best illustrated by their absence, as with the supervisor who "on paper was brilliant" but in the field less so. They were promoted because of plentiful experience in horticulture, but during the season it became apparent their "people skills were non-existent". They could be heard 50 metres away "screaming at someone" who did something wrong or worked too slowly. In contrast, a less knowledgeable grower proved a better supervisor because they ask rather than tell – people skills.

Contrasts between these supervisors reminded me of a training I observed where a specialist in managing horticultural workers advised managers on how to "get the best out of their teams". The first thing a new recruit should see is a smile, he said. His advice foregrounded listening, seeking empathy and trying to work with what motivates people and how they learn best. From Marcia's perspective as a fieldworker in her second season on Connor's farm, being a good supervisor hinges on how they relate to the team:

> definitely it would be nice that they are grounded and patient and observant in terms of like, not so much the crops but people, because everyone's different and actually probably they need quite a lot of people skills to know what's going on with each individual and that really helps and affects the team morale, I think, on the field. It is quite important, that it makes a difference to be able to read people and to communicate correctly.

In the fields being treated as individuals was apparent in how fieldworkers were addressed by name, with managers introducing them to me by how many seasons they had done, or some detail about their background.

On a hot, late-summer day, one group of pickers was focused on the potentially demoralising task of sorting through a quite damaged crop of radish to find thousands of good bunches. They were clustered together amongst the rows, most sitting on upturned crates, casually chatting as they worked. Carl the crop manager led me over calling out "how's everyone doing?" The response from the team leader dripped with sarcasm "Living the dream". But everyone kept going, pulling up plants, looking them over, discarding any too large or small, bunching up six or seven good ones to make the right weight. I recalled what supervisor Jess told me about getting through these tougher jobs:

even if it's you know, not great conditions or you're in pain, everybody pulls each other through and makes the world of difference. The world of difference. It really does, yeah. So I think that's kind of what drives people as well.

She and her team like their jobs because they like each other; they have formed friendships and spend time together outside work. This camaraderie keeps spirits up during tough or boring jobs, so when weeding a field, although heads focus down on the weeds, everyone is still nearby "so there'll be stories and laughter". The second time I saw Jess, this spirit was stretched to its limit, after a difficult season heavy with overtime, too much work and too few people. I'd arrived as her team were taking a break from weeding, a less than thrilling task made worse by having been left too late, and the slopes they had to plod up and down. The team were all smiley and friendly; one had pulled out headphones to chat during break time. Another was being ribbed about being lucky to join another part of the business in a couple of weeks. Jess told me he'd be missed, being good for "lifting the spirits" when team morale dips. This was particularly valuable through this, the toughest season she'd known. Although sympathetic that the work needed doing, Jess was not happy with how things had been communicated from above, or how overtime was determined. "We get put between a rock and a hard place" as supervisors, she said, because they know the orders must get picked, "but then people want to go home". Sometimes people, Jess included, work longer because they "don't want to let others down", meaning the other fieldworkers – their friends.

To get out from the rock and hard place, all that supervisors have are their people skills, in particular communication. But management know they don't always get this right. It takes time to explain why the plastic has to be laid in a certain way in the polytunnel, or what happens to the discarded radishes. In such a large, complex operation, there are many things to coordinate to get good produce to customers, and all parts of the business must align their responsibilities, but like Tom, they are under pressure and do not always have time to communicate more than essential instructions. "Good supervisors" use short weekly meetings to convey information about wider factors, but this doesn't always happen as supervisors are left to do it their own way with limited training.

Connor's boss Logan knows they could do better with professional development for supervisors, but in the absence of this they try to ensure those selected for the role are a "people person". As with many skills essential to horticultural production, people skills are often assumed to be innate, so one supervisor shouts and screams whilst another explains patiently – it is just who they are. Connor was described by colleagues as a good communicator. On one visit I received word he would be a little late, so waited in reception. He arrived with an apology, explaining that a delegation of fieldworkers had turned up first thing with an issue they wanted resolved. He wanted to deal with it straight away so it wouldn't fester. I suggested it was a good sign they felt able to approach a manager in this way. Connor cultivated this approachability by taking time every week to visit teams in the polytunnels, to make sure everyone knows who he is and has some relationship with him. Rather than relying on supervisors to act as intermediaries, he wants the team to feel able to raise issues directly. This is one more thing to do, but it's important, and part of why colleagues consider him "a natural communicator". Some people are good people people, and this is regarded as something you are rather than something developed through practice and training. Coordinating teams of people, keeping them motivated, even when doing monotonous or physically demanding tasks is a key challenge for larger growers, making skills like supervision an essential part of their growing. Despite his moans about the weather, or customers, this is the biggest challenge Logan faces: "as difficult as the vegetables are to grow, it's the human beings who are even more complicated". So people skills are vital.

The skills ecosystem of larger growers

Larger growers include businesses with multiple production sites, 100+ paid workers, supplying large national retailers such as major supermarket chains, or nationwide household deliveries like the box scheme Logan grows for. In addition to the fields, polytunnels and glasshouses where crops are grown, many farms host pack-houses where produce is prepared and packaged. These include companies with overseas operations, a dimension not detailed here as the ecosystem focuses on one farm or growing site within multi-site operations. Many are highly specialised businesses, focusing on a crop or groups of crops with related markets or production systems, for example soft fruits or salad vegetables. These growers are more mechanised and automated, representing considerable capital investment. However, none have yet been able to replace the human labour required for harvesting, meaning most employ considerable numbers of seasonal workers, largely from overseas. The strawberry grower which employs Champion Pickers represents this archetype, as do the cauliflower and salad specialists featured earlier. Most production is conventional agriculture using chemical inputs, although some have pockets of organic growing within their business. Large organic growers like Connor and Logan's employer lean towards this archetype as they employ hundreds of workers with highly structured division of labour, although they connect into certain aspects of the other skills ecosystem, for example through participating in OGA events.

As apparent in Figure 7.3 and Table 7.3, essential skills in larger businesses are distributed across many workers and flow through a hierarchical organisational structure. In comparison with smaller growers, there is greater specialisation and division of labour, including functions outsourced to crop raisers who supply plug plants, and IPM companies providing pest control. Communication and coordination between parts of the business are vital to maintaining complex processes and logistics which get quality produce to

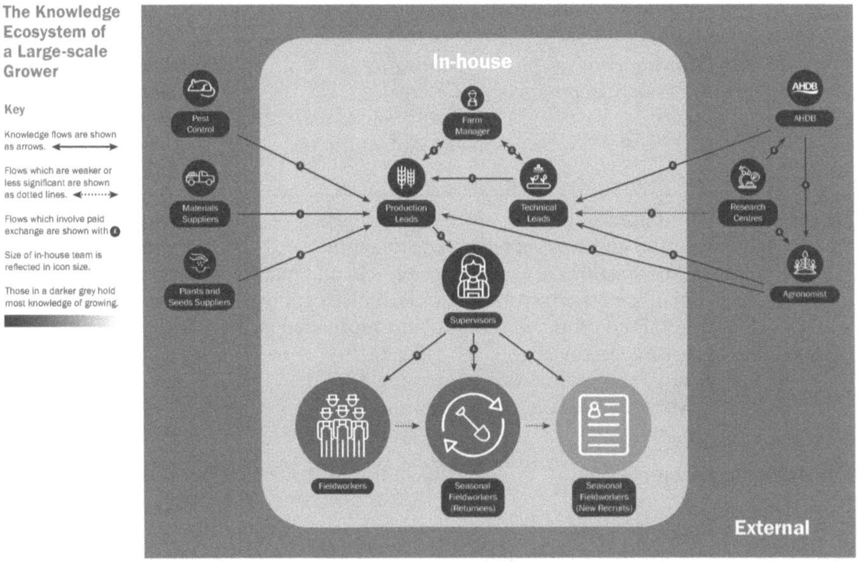

Figure 7.3 Larger growers skills ecosystem.

retailers at the right time. This structure creates opportunities for career progression which can help retain "good workers" and the knowledge they have gained through experience. Distinguishing features of this ecosystem are, firstly, knowledge flows are primarily associated with financial transactions through being paid to deliver a job or service. Secondly, knowledge is highly individualised, meaning expertise is held within and specific to a company with fewer exchanges between growers. Highly specialised operations require expertise which few peers can offer, particularly within the domestic context, and commercial confidentiality reduces the appetite to exchange. Finally, the greatest number of knowledgeable actors and volume of skilled activity occurs at the bottom of the organisational hierarchy, amongst seasonal workers who have the most eyes and hands on the crops. The size of these teams, and nature of the work, makes managing human resources a considerable task and "people skills" a significant area of expertise.

Some contrasts with the smaller grower skills ecosystem are notable. Firstly, there is greater use of data and record keeping to exchange knowledge, making IT systems' materials and data crucial actors. Secondly, these growers employ more individuals with specialist education, for example in agronomy. Where there is no need or capacity to employ a specialist, similar expertise is purchased from external providers. Finally, these businesses are large enough with sufficient economic capacity to support teams or roles dedicated to knowledge development processes such as workforce training and human resources. But there are key ways in which they are deficient in comparison with smaller growers: conventional growers are less able to attract workers by promoting jobs as

Table 7.3 Larger growers skills ecosystem

	Growing Food *Know-how to produce plants*	***Managing the Enterprise*** *Know-how to maintain site & operations*	***Research & Development*** *Know-how to enhance productivity and minimise risk*
Description	Combination of in-house and contracted expertise/inputs (e.g. plants, pest control) Specialised crop-specific knowledge A large proportion of work focused on harvesting	Complex logistics coordinating crop, employees and market People skills keeping teams motivated and efficient Planning and monitoring to maintain efficiency	R&D trials in partnership with researchers Knowledge exchange with crop specialists
Key Actors	Machinery and technology Large teams of seasonal workers and supervisors Smaller permanent teams Specialist and managerial roles (e.g. production manager) Suppliers	Back-office teams (e.g. HR)	In-house specialists (e.g. agronomist) Researchers in UK universities International experts
Knowledge Flows	Learning by doing the job Learning from colleagues Technical education and training (e.g. agricultural degree) Recording cropping plans and outcomes	Internal communications Administrative processes	Recruiting experts from other growers Purchasing consultant advice Delivering crop trials

doing good for society and the environment. Division of labour also means that many roles entail considerable task repetition. This presents opportunities to develop proficiency, for example becoming a champion picker, but can be boring and demotivating, hence the importance of effective supervision or motivation through piece rates. Some large growers mindful of the need to attract the next generation of employees seek to lever social and environmental values as an attractor and consider how to make hands-on roles more appealing, for example through family-friendly hours.

As apparent in Figure 7.3, these growers are large enough to have most of the expertise they need within their workforce. Experienced specialists might have been recruited internationally, or from other UK growers, but specialisation and consolidation mean few high-level experts have experience of each

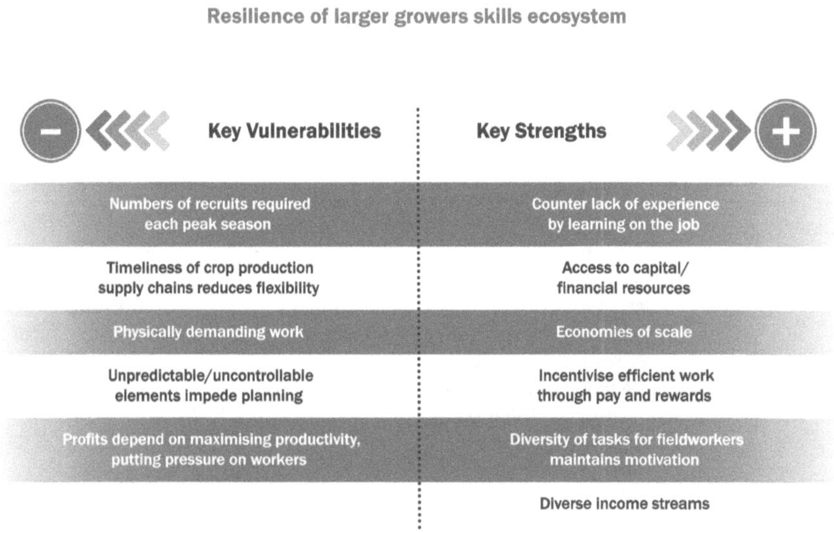

Figure 7.4 Resilience of larger growers skills ecosystem.

crop type (Menary et al. 2019). Larger growers effectively become an authority on the crops they specialise in, knowing more than generalist advisors. As with the smaller grower knowledge ecosystem, many knowledge flows happen through learning by showing and doing. In both specialist technical roles and fieldwork, learning on the job from colleagues is crucial, so knowledge circulates internally rather than flowing into the organisation from outside. Supervisors induct and upskill new starters, then workers repeat tasks, develop their own techniques and coach each other. For more senior levels and specialist roles, learning on the job is prevalent as too few people graduate with relevant qualifications, and those bringing experience from other growers are unlikely to know specific crops or production systems. Specialist advisors also supplement their training and learning from published research with, as one put it: "having the real-life experience" (Sh4).

Growers at this scale experience a range of skills-related challenges (Table 7.4, Figure 7.4), the most prominent being recruitment. Their operations require large numbers of seasonal fieldworkers which have become harder to recruit. These shortages create uncertainty and hamper expansion: "I could quadruple the veg business tomorrow if I knew we had labour" (L4.1). Scale makes this acute:

> it's reliant on bringing in a lot of labour during that period. So there's a big emphasis on: how do you recruit skills, how do you get those skills up to speed as quick as possible and then how can you then retain the skills? (Sh2)

Table 7.4 Knowledge flows in larger growers ecosystems

	Growing Food *Know-how to produce plants*	***Managing the Enterprise*** *Know-how to maintain site & operations*	***Research & Development*** *Know-how to enhance productivity and minimise risk*
Knowledge Impediments	Uncertainty & uncontrollable variables, e.g. weather Poor retention or return rates reduce efficiency and learning on the job A large number of seasonal roles Inadequate supervisory skills	Lack of time and skills for communication Uncontrollable variables, e.g. customer preferences	Lack of time to engage High degree of crop specialisation Unpredictable future
	Importance of timeliness and pace Communicating across languages and cultural groups Working under physical and mental strain Specialisation reduces the relevance of external expertise		
Knowledge Enablers	Motivated workers with the right attitude Good employment conditions High retention and return rates Resources to buy expertise Business scale brings resilience Internal career progression opportunities		

Three important repercussions of this ripple across the knowledge ecosystem: firstly, the labour supply is uncertain making it harder to plan or coordinate logistics in response to short-term changes such as weather. Secondly, temporary roles inhibit learning on the job if workers are not in post long enough to become proficient. A high rate of returnees makes upskilling easier:

> the retention thing, it's just so important because you retain that kind of skills base that you can then use to help, because that's essentially how – through our totally informal training structure – that's how we get the message across about how to do all these quite technically skilled jobs.
>
> (L1.1)

Thirdly, retention and returnees are financially advantageous, reducing the costs of finding and establishing new starters. Experienced workers are more productive and so more profitable, meaning more resources to invest in knowledge development across the business. Profit and skills development are interdependent (Dalizel 2017).

Larger growers have preferred routes for filling seasonal vacancies, with preference being good workers from previous years bringing back skills acquired on the job. For new recruits, employers seek indication that someone will commit to the season, meaning signs of endurance to physical, outdoor work. "Back breaking" work outdoors in any weather or in hot humid polytunnels is not very appealing to many people: "it's just very physical, you can't do it for very long" (L1.1). Seasonal workers do long days, perhaps on a repetitive task which presents another challenge: "Keeping them motivated and happy which is pretty difficult 'cos it's an awful job" (L3.1). It is the supervisors' job to maintain productivity, but they do not always have Connor's people skills, with many having worked their way up based on being a good picker. Effective communication is tricky in a time-pressed setting and is complicated in a multilingual seasonal workforce.

As demonstrated in Chapter 5, growers always face uncertainty around unpredictable variables affecting crops. Weather and plant health have always been unruly, but since 2016 and the UK's vote to exit, the EU growers experience significant uncertainty about staffing which exacerbates other concerns. Worker availability became less predictable and workers cannot be found at short notice: "that lead-in time is now not days, it's a minimum of six weeks" (L3.1). Contingency planning is not always possible:

> people are busy and in fresh produce, time doesn't stop, like you have to just keep going and you have things that are now suddenly urgent.
>
> (L2.4)

Growers have to coordinate the crop's response to growing conditions, with customers' timescales, all in return for a low price, driven by retailer's demand for low-cost products (Medland 2021; Rogaly 2008). It is not hard to see how these intense, pressured conditions sometimes result in worker exploitation; poor treatment and the risk of modern slavery are the nadir of food retail's race to the bottom. Impacts on exploited workers are clearly the prime concern. There are also knock-on effects for the industry as a reputation for exploitative practices makes it harder to recruit or to secure public support.

As highlighted in Chapter 3, the knowledge ecosystem around larger growers has been depleted through reduced public investment in R&D and extension. This is the opposite of the supportive environment necessary for a healthy skills ecosystem (Brown 2022). Smaller growers share this national context, but organic and agro-ecological growers have always been less engaged with state-supported agricultural institutions. They complained that organisations like AHDB did not research innovations relevant to them, instead focusing on conventional agriculture. Conventional growers and grower groups do invest in crop research, but tailor projects to their needs. A consultant working on trials explained they benefit fewer growers because findings are retained for those funding the trial. In contrast, levy funded R&D had knowledge exchange "built in", meaning learning was disseminated more widely (Sh3). When this consultant runs a knowledge exchange activity, she tends to get the same growers attending, suggesting that only larger growers have the capacity to devote to R&D. The consultants themselves, employed by a profit-oriented company, struggle to keep up with new developments in the field due to lack of capacity. R&D investment has become increasingly individualised, financed only by actors with sufficient economic resources, who then retain the benefits.

In face of these challenges, growers at the larger scale have features which help ensure access to knowledge and skilled workers. Firstly, they invest in conditions which enhance retention rates such as quality accommodation and financial incentives. Their economic capacity is "fuel" to feed the ecosystem (Brown 2022). One grower said annually they pay thousands to enhance on-site accommodation, providing leisure facilities and pastoral care; in return their reputation as a good employer attracts returnees, their friends and family. Secondly, growers are maximising productivity through initiatives like Champion Picker training and the use of LEAN. Thirdly, growers are evolving production to adapt to workforce availability, for example creating part-time family-friendly roles. Where businesses produce multiple crop types or combine horticulture with other production, they might offer year-round posts attracting local recruits:

> So you're trying to set these bits of the business up to have permanent staff year round [...] you work from the labour backwards.
>
> (L4.1)

Actors in the knowledge ecosystem are applying corrective mechanisms, adapting to retain resilience (Brown 2022). Larger growers can make such enhancements because they have human resources specialists and financial capacity to invest and await payback; they are economically robust enough to ride out short-term challenges. The health of this skills ecosystem is therefore dependent on their profitability, which is determined by the agri-food regime they produce for, and which tends to encourage scaling-up (Chapter 1).

The health of skills ecosystems: Two paradoxes

Looking in detail at the skills ecosystems of growers at different scales indicates sector-wide tendencies which are apparently contradictory, and which are shaped by the meso and macro operating contexts. Firstly, the work is hard and relentless, but can be enjoyable and rewarding for those who like being outdoors. These qualities are why people become horticultural workers, whilst deterring many others. Secondly, even in the more sharing-oriented ecosystem, the risk of skill flows being blocked or depleted sits with individual enterprises. Smaller growers' strong networks and collective activity are reliant on individual growers surviving their vulnerable positions. Conversely, larger growers are individually relatively strong but demonstrate weaker collectivity, meaning risks around skills sit with individual growers. Each skills ecosystem is only as healthy as the individual growers within it. Next, I consider these paradoxical traits in greater detail.

The first paradox is that work across horticulture's skills ecosystems is "back breaking but spirit lifting", as seen in the trope of seasonal workers "hard but rewarding work". A description from the lynchpin of a family farm encapsulates how the work breaks backs but lifts spirits, particularly as peak season hits:

> There's a point in March where you certainly notice that from walking [past] around the farm you're running around the farm. But there's also a point where my spirits tend to lift with that because it's generally – hopefully – in May getting warmer, so lovely bright mornings, often. It's more appealing going outside than it was a month previous to that so it gets easier. So there are those emotional factors as well.
>
> (S1.1)

At any scale, producing fruits and vegetables can be "back breaking". In some roles "back breaking" work is less literal, referring to the workload and stress of managing the vagaries of growing crops, but other roles literally strain backs, through bending and lifting, again and again. In combination with wet, cold, muddy, outdoor conditions, this takes its toll:

> The most challenging? Maybe the winters. The wet days, the shit days. I would say yeah, just like keeping motivated, I guess
>
> (S1.3)

It makes you wonder why anyone would do it, particularly given the economic pressures and social constructions which have devalued horticultural work globally. But across the industry, I never got the sense that people do these jobs for want of any other work. The exception may be those on seasonal visas tied to the sector, or migrants who lack language skills for other low-pay sectors such as hospitality.

Many people actually want to work in horticulture, a choice with three common motivations. First are those from farming backgrounds who always knew they would go into farming. Many large UK growers began as family farms and still think of themselves as family businesses, run by fathers, sons and daughters. Some individuals work with these companies to gather expertise to take back to their own family farm. But a vocation in food production is not always inherited and can be driven by the industry's importance:

> you're the one picking the food, supplying to all of these families across the UK, and it's good food so it's a good feeling that someone's going to be eating good stuff.
>
> (L1.5)

One graduate working for a large grower saw himself contributing to "feeding our nations" which is challenging but:

> it feels quite a noble, exciting thing to be doing. It just feels, you know, like you're doing a lot of good in the world, I guess.
>
> (L2.3)

Similar notions drive many working in agro-ecological production who see it as a values-driven role (LWA 2024; Soil Association 2018).

A second set of motivations connect to the desire to grow plants. Horticultural work can be rewarding, as evidenced by the legions of hobbyists who do it for fun and relaxation (Pitt 2014). This feeling is not lost at the commercial scale as this plant raiser suggested:

> So it's quite nice to know that you've produced a product from seed and people have got that on their plate like, you know.
>
> (Su1)

Years of toil and millions of plants have not dulled his excitement and satisfaction. Other features celebrated by home gardeners motivate those in hands-on roles:

> Just being outside, working physically, it's quite important for me. Yeah, I love physical work. It just makes me feel better and just generally, I feel like it's quite important, like growing food and working in that industry, something that everyone needs. It just feels like quite a wholesome job, yeah, quite important.
>
> (L1.6)

The values and nature of the work come together in a particular way of life, which appeals to volunteers seeking experiences rather than pay. Take this volunteer's response to my request to describe his role:

> I'd describe it as a really fulfilling way of occupying yourself. You're not just in a factory production line doing the same thing. You've got a bit of creativity and a bit of innovation within the job and I just really, really, really like the lifestyle. I like the rural lifestyle where you actually wake up with the sun and then go to bed feeling tired but in a good way, like you actually sleep well … Because you've been working, and I really like this part of it because it's a lifestyle thing and then it makes sense, this kind of rhythm and, yeah, being a farmhand that makes you think like you're a helper but also you're bringing something to it, as well.
>
> (S1.4)

As we've seen, this lifestyle is not all "skipping around a meadow and picking daisies" as some people imagine (L1.4), but its appeal can see people through the wet, shit days. A farm manager who was fairly downbeat about logistics and worries, such as drought, talked about them in a manner so pessimistic I struggled to engage with him. Then, as a day driving round crops wound to its end, he slowed the pick-up at the top of a sloping field, revealing a beautiful landscape extending for miles, shining in the summer sun. He turned off the engine, we both looked out the windscreen, and he told me some days he just drives up there to enjoy the view, taking a moment. Sometimes the rural is actually a little idyllic.

The final set of motivations are financial: although having a reputation for low pay, particularly considering how hard the work can be, horticulture can be financially rewarding, and not just for managerial roles. Those who recruit fieldworkers identified this as a key attractor:

> So, my teams are generally reasonably unskilled as staff, they're normally very often quite bright people. I would say you know, there's no flies on any of them to be honest, they just have chosen not to excel or work in these sort of ways and their motivation is driven in a much different way than other people are. They are pretty driven by money, over a time the prospect of going on a salary, they'd find horrific.
>
> (L4.2)

This team mixed seasonal migrants and local recruits drawn to outdoor, physical work; piecework means all earn well despite not having advanced education or qualifications. Combinations of these rewards and motivations mean many people find it worth enduring the tough parts, finding a balance between breaking backs and lifting spirits – or bank balances – which ensures horticulture offers good jobs for the right workers.

The second paradox arises from relationships between growers and their collective responsibility for skills ecosystems. Resilience of both knowledge ecosystem archetypes, despite having markedly different degrees of knowledge sharing, depends on individual businesses' economic viability. In the ecosystem of smaller growers, knowledge is readily shared amongst a vibrant community of practice working to educate future growers and exchange learning, advocating for public investment in knowledge exchange. The community of agroecological growers treat knowledge as a common good to be distributed for collective benefit, rather than individual property serving individual needs (Anderson et al. 2019). But the ecosystem functions to a large degree on goodwill and without direct financial remuneration for expertise contributed. This is a strength in that the flows are independent and self-sufficient, a system which arose because public provision was lacking. But this is also a weakness given the economic vulnerability of many smaller horticultural businesses. Growers cannot always afford to donate time to knowledge exchange, plus there is a very real threat of going out of business, meaning on-farm learning opportunities would disappear. The risk of knowledge flows being blocked or depleted still sits with individual enterprises: if too few smaller growers remain able to share knowledge, then the pipeline developing future growers becomes blocked. Individual trainees bear the risks of low or unpaid traineeships, and training lacking guaranteed quality or conditions. The knowledge ecosystem of smaller growers is only as healthy as the individual growers who feed knowledge flows.

Within the other archetype, the risk is individualised in very different ways. This skills ecosystem is concentrated on resources held by each grower – the skilled people who work for them and their capacity to purchase external expertise. Since the demise of publicly funded horticultural research, knowledge development centres on needs of the businesses paying for it. If larger growers lack resources to invest in knowledge, there is no one with the motivation or capital to step in. In the hands-on domain of skills development, knowledge is also individualised to the business as training is largely in-house and few workers bring relevant qualifications. Even graduates with specialist education require on-the-job training to know the company's specific crops and operations. The relative lack of exchange between larger growers, with far fewer active networks and peer-learning flows than between smaller growers, exacerbates knowledge privatisation, tying skills to individual companies. That is not to say larger growers never exchange or cooperate, but this is less prevalent and seems less essential for them. The demise of levy supported R&D signals the limits of shared responsibilities and action amongst larger growers, further individualising ownership of and investment in horticultural knowledge. If R&D investment flows between company and researcher with no public or collective intermediary, then the knowledge is privatised, preventing access for those developing new businesses or public goods. The result is a less resilient agri-food system focused on commercial interests (Anderson et al. 2019). The sharing-oriented ecosystem is not perfect, but does suggest potential for a food growing knowledge commons – a shared good available to benefit all (Vivero Pol et al. 2020). In the final chapter, I consider how risks currently threatening this ecosystem could be minimised through expanding collective responsibility for knowledge flows.

Conclusion: Restoring ecosystem health

This chapter zoomed out from the micro-level of growers' skilled work to the world in which they operate and learn how to grow. Sketching skills ecosystems elaborates the multi-scalar links between knowledgeable actors, and how systemic pressures impede knowledge flows. Positioning skilled work as a relational achievement shaped by social context counters the individualising tendencies I have criticised across preceding chapters. Drawing analogy with biological ecosystems, I have suggested that flourishing horticultural skills ecosystems are diverse and healthy enough to support knowledge services essential to abundant food growing. Knowledge flows are unimpeded, and all actors have adequate resources to feed their knowledge exchanges; access to the system is open to all, and the ecosystem is resilient in the face of change. At present, horticulture's knowledge ecosystems are vulnerable, dependent on the relative strength of individual growers and their collectivity. Smaller growers are relatively weak as individuals, but have strong networks and connections which bring collective strength. Conversely, larger growers engage less in collective action, but are stronger as individual businesses. For the former, strength comes through connection, whilst for the latter strength comes through scale. Smaller growers are interested in gaining resilience not through individual businesses scaling up, but through cooperation and collaboration. Larger growers might strengthen their skills ecosystem through more collective investment in knowledge flows and by treating their workforce as skilled individuals deserving quality jobs. At both scales, but through different dynamics, the resilience of the knowledge ecosystem depends on the health of individual enterprises which are vulnerable to pressures corporatised global agri-food systems exert on producers. Ensuring the knowledge required to keep growing food in the UK remains available therefore depends on a vibrant horticultural industry in which all types and scale of operation are more profitable.

Devalued horticultural knowledge and skills contribute to these dynamics and are outcomes of them. Products of unskilled work are less socially valued, so horticultural workers' labour is under-rewarded; producing low-price foods cannot support quality jobs and high wages, and low paid workers are assumed to be unskilled. Because the sector's jobs are socially devalued, there is less demand for educational courses, so investment in the knowledge ecosystem declined, leaving it depleted. The sector has compensated through relying on a workforce with little formal specialist education, which is in turn perceived as low-skilled, perpetuating negative feedback loops.

A focus on skills ecosystems suggests horticulture risks depleted knowledge infrastructure. These risks do not arise within specific domains of knowledge development, but from the economic dynamics of cheap food production. Growers operate with low profit margins because prices are squeezed downwards whilst costs keep rising; labour has been the most contractible budget line and so has been squeezed particularly tightly. In this competitive marketplace, it is difficult for growers to invest in knowledge development – profit and skills are connected. Add to this a lack of public funding relative to other types of agriculture, with less income subsidy or state provided education and R&D, and the result is an environment where essential horticultural resources are highly depleted. Horticulture does not face a skills crisis per se, as the causes lie beyond knowledge dynamics, being firmly rooted in the problems of global neoliberal market-driven food systems which make it incredibly difficult to secure a good livelihood from growing good food. Solutions focused on better training or professionalisation will therefore have limited impact. But the systemic drivers of the wider agri-food crisis are more intractable, making it worth considering what can be done to shape healthier knowledge ecosystems *for now*. I suggest this requires moves beyond the individualisation of risk and resources associated with horticultural skills towards ecosystems circulating common knowledge.

8 Conclusion

Revaluing horticultural work and workers together

Sharing tips and thrips: Exchanging knowledge in the field

We all crowded round the table as Greg unscrewed the insect trap and poured the contents onto a large white platter. Out came sticky brown liquid and a number of drowned insects. Hand lenses already out, the entomologists leant over and were soon calling out the names of what they had identified, far too quickly for me to catch, especially when they flipped between common and Latin names. Greg explained he'd put the traps amongst the fruit trees last week, ready for us to inspect today. Having been involved in IPM (Integrated Pest Management) trials, he had been asked to share his experience. On the white tray, we were looking for signs of spotted wing drosophila, a type of fruit fly newly arrived in the UK, bringing the threat of serious crop loss through fruit collapse. Our host held up a large poster produced by AHDB which is on display in the packing shed to help identify the fly, promoting the importance of "good hygiene" around fruit trees to make conditions less favourable. It wasn't long before hand lenses hovered over a tell-tale wing spot; Carolina sounded almost excited as she pointed it out, despite the bad news it brought Greg. Perhaps to compensate, her colleague also pointed out a soldier beetle, "a good guy".

Next to be laid out on the table were several sheets of A4 paper, neatly stuck into one long thread, displaying Greg's equally neat handwritten calendar of pest control measures. Month by month, crop by crop, columns noted measures of biocontrol or chemical control, and the target pest. Other growers crowded back round the table to take photos; Geri who organised the visit asked if she could share with others in the network. Greg developed the plan over the years, incorporating advice from the specialist present that day, and another consultant, creating an IPM system tailored to his crops.

Before the white platter was covered in insect trap debris in this impromptu IPM lab, similar ones served lunch on the same table as the group chatted, passing the food organised by Geri and cups of tea brewed in the house. It wasn't just sandwiches being shared that day. Promotion of the training mentioned sharing several times: growers will share their learning, and specialist advisors will share their expertise. That is exactly what had happened.

DOI: 10.4324/9781003281689-8

Figure 8.1 Sharing insect identification on the picnic table.

I'd arrived that morning to join a huddle of growers from across the region chatting just outside the first host's polytunnel, waiting for late arrivals struggling with rural postcodes and Google maps. Once all names were ticked on Geri's clipboard, we were led through the tunnels and germination rooms by Iwan, who told the story of his journey into growing and how the site operates. Visiting growers chipped in questions as we went: how was it finding staff for picking? Where did the seeds come from? Who are the customers? Can we taste the produce? Topics ranged well beyond IPM which was the designated theme for the visit. Occasionally the IPM advisor prompted Iwan to talk about particular controls, or how he spots signs of different pests, but the conversations were loose and he seemed happy to talk about any aspect of his operation. Some visitors scribbled notes, others took photos; they formed ad hoc pairings walking and talking about their own places, exchanging gripes and victories. These huddles occasionally paused to point and peer at something unusual on a plant, postulating what it might be, perhaps calling over the entomologist to confirm an identification, or to diagnose an unknown sign of disease. They looked for the "little white husks" Iwan had mentioned as signs of aphid presence, something he keeps "a radar" out for during crop walks. First tour over, one of the advisors brought the learning together with a summary of IPM principles and advantages. They concluded that you have to be a "bit relaxed" as there will be pests, so you have to trust that the controls will do their job.

Questioning peers, noticing things about the plants, making informed judgments in response, consulting experts for more guidance – many of the ways growers learn were apparent during this visit. Perhaps most striking, though, was the spirit and practice of sharing. Gowers invited other growers to see and hear the detail of their operations; the specialist advisors were there to contribute expertise. They went home with 300 thrips collected from the polytunnels to test back at the lab for insecticide resistance, adding to the research which in turn will be available to other growers. Carolina, the thrip collector, later told me that this type of study visit is a highlight: "The best part of the job, yeah, I think that has got to be engaging with growers at events". Days like this are good because "people are pleased to see you". She liked that they were engaged and found "it's a good chance to show off everything I've been learning as well – it's quite a rewarding experience".

During the day Carolina shared tips on spotting and identifying thrips, pointing the insects out on plants and answering queries about controlling them. This type of visit, seeing things directly and talking to other growers, is her favoured way to spread expertise. Trials and new information are written into "these very lengthy reports" publicly available online, but Carolina is skeptical about their value: "I wonder how many growers end up reading those". In-person events seem much more effective for knowledge exchange: "you get a lot more interaction directly from the growers asking questions, and they can talk to each other as well". Her colleague added that people seemed to have missed this type of gathering during the pandemic, so the return of site visits was well received as a chance to catch up: "Just in that, 'Oh, hello, I didn't expect to see you, haven't seen you for ages,' and the general sort of learning there, peer to peer style". Visiting other growers seems to work for hosts, visitors and specialists who learn from seeing the growing in its real context: "it's a two-way street".

This visit was coordinated by a project funded to support growers with activities and networks responding to their needs. After a couple of years working on it, Geri the coordinator knew everyone who came, giving each a friendly personal welcome. Such interaction was part of the programme's ethos as she described it: "We are trying to get people working together, and it's that soft skill exchange as well as the specific courses, that hopefully they'll benefit from". One of her team told me that a colleague working with other types of farmers was struck by the difference in the horticultural groups – much more open, more sharing. But this spirit hadn't been a given: "When we started we were probably told 'oh you won't get the growers working together'". This is a stereotype I often heard from people across and around the sector. But in practice, Geri had not found it the case; they watch for commercial sensitivity and try to be responsive to growers rather than dictating to them. It seems to be working. Greg and Iwan were paid for the time and trouble of hosting the group and received a round of applause at the end of the day. Carolina had her thrips. The other growers went home with ideas, perhaps some new phone numbers, lips and teeth still red with the juice of fruit plucked and tasted along the way.

De-individualising horticultural skills

This vignette highlights growers sharing knowledge, and activities designed to foster skills benefitting the sector. I opened with it because the spirit of collective action contrasts with tendencies apparent across horticulture's skills challenges and how they are imagined. Across this book I have highlighted how assumptions about horticultural skills and work tend to individualise them, locating problems and solutions with individual people or businesses, neglecting their systemic roots and connections. In these flawed narratives, horticulture's labour force issues are associated with individual workers, not the work: a hard-working migrant simply has the will and endurance to spend a season picking fruit, whilst a lazy local doesn't. A grower who can't attract enough recruits is responsible for improving pay and more attractive offers; when employees are exploited, it is the fault of a rogue recruiter. The same tendency is seen in thinking about the (un)skilled nature of horticulture work, which contributes to its undervaluing. Where a horticultural worker is recognised to have important knowledge and abilities, their skill is regarded as something a body has or not: you were either born with the knack or not. Some people are just people people. You have the high skill level to secure a visa which opens the door to immigration, or you can't have the key. Desirable skills are qualities associated with the individual, making shortages a problem of too few individuals holding or developing horticultural skills. The problem is not constrained career opportunities or inadequate educational infrastructure, or that the grower's lack economic capacity to invest in professional development. And solutions are left to each grower to fathom.

Through these flawed but influential interpretations, horticulture's skill crisis has been made a matter for and of individuals, with skill a free-floating capital which can attach to them, or not. Skill has become a fetishised commodity, neglecting the relationships and practices through which it exists and is learnt, and systemic pressures which deplete or impede its flows. Rather than seeing skill as practices developed in social contexts, it is assumed to be something individuals have or do not have; responsibility for acquiring the skills which gain someone the right to immigrate lies with the individual. Learning on the job is under-appreciated as a source of specialist expertise, feeding the belief that horticultural work requires practical ability more than specialist knowledge. Workers are valued for the number of labour hours they bring, rather than the quality of their abilities, or through being offered good livelihoods in return. Seasonal workers are treated as free-floating individuals separated from their social relations and lives.

This individualisation rests on false interpretations of how skills and work work, separating mind and body, with intelligence located only in the former, relegating practical, physical roles down the job hierarchy. It also distracts from and obscures the real problems with horticultural skills, and horticulture in general. Systemic problems of unjust agri-food regimes and wider systems of capitalism, racism and patriarchy which they are embedded in are neglected through a focus on moving adequate numbers of workers into place. Debates focus on traits of archetypal characters, rather than considering the context shaping employment practices, and agri-food systems which shape labour regimes requiring flexibility and creating precarity. When skill is fetishised, it is too easy to assume that good growers are born not made, obscuring the learning it takes to become a grower, and ecosystem properties which allow skills to thrive. Investment in innovation and research is left to individual companies to fund and benefit from, making that knowledge private in both senses of the word. Causes of and solutions to skills crisis are positioned with individuals, to the neglect of systemic factors, or collective responsibility to shape economies in which good growers are fairly rewarded and treated.

Contrary to myths and histories imagining horticultural work, I have shown it entails complex and comprehensive skill sets – the knowledge to grow food, run an enterprise and adapt to change. Good growers attend to multiple variables by looking inward to the plant, and outward to the world they sell into, then judge how to respond. Horticultural knowledge extends throughout all levels of the work, enrolling things and nonhumans, from pencils to apps. Tricky skills of bending plant lives, whilst never being wholly able to know what they are up to, are crucial, unique dimensions of growers' know-how. Far from being individuals with some innate knack, good growers have taken time and effort to learn specialist know-how rooted in experience, learning which is thoroughly social as novices and experts show and do together. But knowledge flows become blocked or are inaccessible to some; the ability to develop horticultural know-how depends on access to places to become a grower, which takes land, capital and agency.

The skills ecosystems which provide both larger and smaller growers the knowledge resources they need to thrive are vulnerable in different ways, but in both cases, risks lie with individual growers and the health of the ecosystem rests on their viability. In the more sharing-oriented ecosystem of smaller and agro-ecological growers, flows risk being blocked and depleted when individual growers burn out or go bust. Each larger grower is relatively resilient, but with fewer links between them and little collective responsibility for knowledge, their ecosystem also depends on individual businesses. The skills ecosystems are only as healthy as the individual growers they comprise, and their health is determined by economic viability. For UK horticulture squeezed by global food systems which treat fresh produce as a cheap commodity, this is deeply problematic.

Although I hope that my attention to horticultures' skilled work demonstrates that these workers deserve to have their roles reconsidered and revalued, it is insufficient to regrade their jobs as highly skilled. Any such categorisation risks perpetuating deeply problematic constructions of people and their work, drawing on racialised and gendered notions of who deserve status as valued workers and citizens. It also diverts attention from issues really preventing workers from attaining fulfillment and fair reward: the ownership and organisation of their labour. Reskilling is about more than understanding whether jobs are complex and intelligent, and must be approached through interrogating contexts in which people work, including the power dynamics. To truly revalue horticultural workers means giving them a stake in their work and its products, allowing them greater control of their labour and how it is rewarded.

Horticultural work and skills have been devalued through systems and thinking which individualise the problem and the workers, so I suggest that to restore the value they deserve requires the contrary: collective action and responsibility. Maintaining flourishing systems of horticultural knowledge is not something individual growers can do, nor should they. A healthy ecosystem of horticultural knowledge is a common good and we are all responsible for providing resources to enable it to flourish. A better way to imagine and work with horticultural skills starts by treating it as common knowledge, with benefits and risks shared amongst a community which also shares responsibility for ensuring healthy knowledge ecosystems. To suggest what this might look like, I draw on the idea of the commons in relation to food and knowledge systems. Then I highlight pockets of practice where growers take an approach more aligned with commoning. Together the theoretical and the actual point towards a knowledge commons which de-individualises horticultural work, so helps revalue it.

Towards a knowledge commons

In Chapter 3, I described UK horticulture shaped by dominant agri-food systems which treat food as a commodity traded for private profit. In an alternative vision, food is a public or common good produced to benefit people and the planet rather than the few who benefit from profitable trading. Food as a commons rather than a commodity shares risks and benefits between the community; everyone needs to eat, so it is our collective responsibility to ensure food production. The most transformative visions of food commons reach beyond ensuring everyone is fed, considering how production is governed through collectively made decisions involving everyone who eats (Vivero-Pol 2020; Vivero-Pol 2017). A food commons is only truly so if production is based on values other than profit, meaning nature and labour involved in production are non-commodified (Pettenati et al. 2019). De-commodified labour is not just a question of lifting the fetishistic veil to reveal the workers behind the products, as the ownership of the means of production also must be revealed, to distribute power and value fairly (Pettenati et al. 2019).

Food as commons suits the goal of revaluing horticultural work, because it considers food as more than commodity, meaning it has value beyond economic profit (Vivero-Pol 2017). This broader value includes that returned to workers, not just as pay exchanged for the value extracted from their labour, but in terms of impacts on their well-being and the esteem they receive. More radical interpretations of food commons, or commoning, look beyond common access to the resource – food for all – to consider social processes and relations around it, and the politics of how they are organised, bringing the potential for food justice. Collective ownership and action can counter injustices in dominant food systems, for example through alternative exchange mechanisms such as cooperatives (Cadieux and Slocum 2016). Cooperative ownership by food workers resists capitalist markets by distributing wealth horizontally instead of up to business owners, so labour contributes to communal rather than individual benefit (Alkon 2014). I suggest such structures also positively impact knowledge ecosystems.

A commons approach has been explored with a specific focus on knowledge in agri-food systems, drawing contrast with "traditional" agricultural knowledge in communities where it is shared and self-governed (Reyes-Garcia et al. 2020). Anderson (2019) argues a moral imperative to treat scientific knowledge used in agriculture as a common good because it is essential to ensuring the universal right to food. But as I highlighted in relation to horticultural R&D in the UK, agricultural research has become highly privatised, limiting access to those with funds to invest or to purchase the outputs (Glenna 2020). Global neoliberalism tends to reduce or privatise public extension services, whilst urging smallholders to rely on commodified inputs, including patented products in which corporations capture communities' traditional knowledge (Zerbe 2019). But any science depends on the knowledge which preceded it, and thrives through the free exchange of information, so is really a shared endeavour and global common good (Anderson 2019). Agricultural knowledge cannot be privatised or owned because it builds on knowledge already in the commons, so should be regarded as a common good.

Unlike other shared resources, knowledge is not exhaustible or limited as one can pass it on whilst still retaining it (Helfrich and Haas 2009). It suits commoning because "one person consuming knowledge doesn't diminish it for others" (Anderson 2019, 186). Privatised knowledge falsely constructs scarcity and exclusivity through copyrights and patents which treat it as rivalrous and excludable – only some can have it (Helfrich and Haas 2009). But outside of market dynamics shared knowledge is not depleted. I would go further to say that commoning the knowledge required to grow food actually increases the knowledge base. When a grower teaches a novice how to grow crops, the outcome is two knowledgeable growers, each of whom might go on to share what they know with others, expanding the reach of the know-how and strengthening the knowledge ecosystem.

To de-individualise horticultural skills, what inspiration can we take from knowledge commons and commoning? In knowledge commons, ownership, responsibility and benefits are shared, making it about community and how relationships are governed (Helfrich and Haas 2009). Forms of relating are crucial, as resources only become commons through cooperation (Caffentzis and Federici 2014). Commons do not share "things" like commodities, but shape social systems and practices (De Angelis 2014). Some interpretations focus on shaping institutions and rules to govern the resource and community (Dedeurwaerdere et al. 2014). Others celebrate experimental ways of being through interdependence and shared vulnerability, in pursuit of better living through connecting with others, including more-than-humans (García-López et al. 2021). Commoning attends to qualities of relationship, seeking care, cooperation and responsibility, in a community of those sharing responsibility and entitlement (Caffentzis and Federici 2014; Federici 2011).

The community sharing responsibility for horticultural work and skills extends beyond growers and others directly involved in production to those who benefit from what they know and do; those eating the produce have responsibility to care for the natural and social resources used to produce it. The common resource is more than the know-how applied to grow food, because this knowledge cannot be abstracted from production practices, both because of the significance of learning by doing and because of the need to adapt to particular crops, places and times. It can only be shared in practice, in the learnings and doings of horticultural work as they happen in colleges and fields. So a crucial element of this commoning must be equal access to the ability to produce food, a central right within food sovereignty (Holt-Gimenez and van Lammeren 2020). This clearly contrasts with current tendencies for markets to squeeze producers out of growing or to prevent their entry. The connection between knowing and doing food growing suggests that a commons which de-individualises the challenges I have identified is not constrained to knowledge dimensions of food production. They must challenge dominant agri-food systems and inequalities which shape them, including ownership of production.

The goal of the commons is to create alternatives to capitalism (Caffentzis and Federici 2014). No small task, and seemingly idealistic, but seeds of these alternatives exist, across history and in spaces of resistance living against capitalism's apparent hegemony (Caffentzis and Federici 2014; Gibson-Graham 2006). Communities save and share their own seeds in spite of international regulations which seek to enclose knowledge and natural resources vital to agriculture (Sievers-Glotzbach et al. 2021). Peasant seed groups collaborate to breed resilient crop varieties, developing a collective knowledge base grounded in collaboration and exchange (Mazé et al. 2021). These and other examples of commoning strive to maintain spaces free from market and state control, demonstrating that human life is never wholly driven by economic rationale (Caffentzis and Federici 2014; de Angelis 2014).

Seeds of alternatives to capitalist food systems, closer to commoning than commodity, have sprouted across the horticultural systems I have explored. Growers sharing knowledge with each other and collaborating to promote a shared vision for better food are sowing seeds of resistance alongside their crops. Sometimes they struggle to keep these seedlings protected from the weeds and pests of low-price production regimes which too easily shade or infect them. As one grower at a recent workshop on agro-ecological training reflected, it's hard to imagine ideal scenarios when the world isn't ideal. For now, they seek solutions "in the meantime" – achievable improvements which might step towards bigger change, not necessarily radical, but hopeful. I want to highlight examples I have encountered which point towards revalued horticultural work, through collective activity within more just food systems. They are not perfect, and do not represent fully realised commons, but are steps towards commoning which help understand how it might be achieved.

Growing collectively – "Power with"

The experiences and perspectives I have shared so far indicate currents within horticulture in the UK and globally, which challenge the imperative to individualise and commodify food and knowledge. Despite stereotypes about their businesses, and pressures pushing in the other direction, some growers collaborate on shared interests and favour cooperation (Nicol and Taherzadeh 2020). Smaller growers know they are vulnerable to the challenge of balancing workload and profitability; they lack efficiencies and resources which come with scale. But none of those who discussed this suggested the solution was to get bigger, rather they emphasised the need for more cooperation and collaboration to increase individual businesses' resilience. They want there to be more growers like them, hence committing time to share what they know, supporting growers who will work alongside them to produce more, better food. Beyond knowledge sharing, some aspire to collaborate on producing and selling. Tom would love time to develop connections with other organic growers to form a cooperative helping him out of his labour-profit squeeze. Each grower could focus on what they produce well, so everyone's customers can get a good variety, without everyone having to do everything. Such division of labour would gain advantages of specialisation and scale which larger growers have, whilst retaining their human scale operations and values-driven systems. Auction house models based on the system well established in the Netherlands have also been suggested to enhance smaller growers' economic position within supply chains (Tourte and Gaskell 2004).

The value of cooperation is recognised in other pockets of UK horticulture, for example amongst larger growers concerned with the risks of losing the statutory levy as a collective R&D investment. They have been advocating for a continued but improved levy to fund research and innovation, as "a system of co-operation for the benefit of all growers" (Growers Better Levy Group n.d.). Despite being a sector evolving at pace, where ideas are currency, I also encountered a spirit of sharing and collaboration within CEA. The company visited in Chapter 5 aims to produce systems which can be repaired by growers or by their local plumber and electrician. This is contrary to many tech companies' tendency to retain control over digitised systems, so customers keep buying services and corporations expand their control of food production (Carolan 2020; Klerkx et al. 2019; Rotz et al. 2019). Rather, the company's ethos is about "empowering that grower" through easy to use and learn tech, and publicly available data. Their vision for the evolution of CEA is a sector of connected specialists in a "collaborative ecosystem where everyone finds their niche"; no one company becomes huge and rich but "everyone does quite well". This seems counter-intuitive to a company founded on intellectual property, so what is the thinking: "We want vertical farmers to be a success, I think it takes everyone coming together".

Collaboration has the potential to become more than a seed through increasing growers' power to resist economic pressures or to push for transformation enabling their vision of values-driven food production. Commoning can help communities gain "power with", as individuals amplify their voices to effect change by coming together (Leitheiser et al. 2022). Individually, small growers lack the power to advocate for change or to resist actions by those who hold more influence within the supply chain – the price makers. As coalitions or movements, they become stronger. Like workers who have influenced labour controls through collectivising into unions, growers could effect change through collective action. To date, *power with* has been most effective in relation to horticultural work through larger growers' lobbying for access to seasonal migrant workers. But growers might collaborate to apply *power with* to push for changes with more just outcomes, perhaps combining with consumers to expand their reach across food systems, to benefit everyone who works to grow food (Klassen et al. 2023).

Sharing the means of production

The collaborative spirit highlighted so far demonstrates that horticulture does feature the will to work together to improve growers' situation, but not all examples reach into fields or polytunnels to benefit the majority of workers. Commoning to improve their situation requires us to question who owns labour and the means of food production (Pettenati et al. 2019). Cooperative ownership models can enable commoning by changing workers relations to their labour and its products and by changing wealth distribution. UK horticulture features this in action, with the organic grower Connor works for being wholly employee owned and run to benefit all workers. This change came with a new governance system and organisational values defined through staff consultation. Everyone who contributes is rewarded as co-owners – including seasonal staff – receiving a profit share distributed non-hierarchically, so it is in everyone's interest for the business to do well. This structure conveys that all roles are valuable to the business, from cleaner up to senior manager.

Employee ownership is appreciated for "a greater feeling of oneness" across the organisation, as one put it. Staff feel invested in the organisation: "because we are owned by ourselves we're more interested in looking after ourselves". Everyone shares responsibility for improving conditions, meaning it is in everyone's interest to raise issues through their elected representatives. Those I spoke to were positive that employee ownership promotes a sense of engagement in which people are involved and feel part of *their* company. Staff were quick to speak about enjoying their work and a "family feel" where "everyone actually really cares". Recruiters find that "people do actually want to come and work here", particularly when their values align with the company's. A newly recruited driver almost turned down a job offer worrying that everything about the employer seemed too good to be true – what's the catch! Once in the company, many want to stay, working up through roles as their good work is recognised: one receptionist started as a cleaner. Movement between teams was seen to indicate a collaborative spirit, long part of the company's ethos, made tangible in co-ownership: "everything we do as individuals on a daily basis whilst we're at work contributes to success, or failing".

Employee ownership places the means of production in common ownership, and in this example comes with governance giving all workers a voice. Employees benefit by sharing profits, and the business attracts a motivated, collaborative workforce. Relationships strive towards values like those sought in commoning, such as care. Work in the fields and polytunnels can still be repetitive and physically demanding, but tasks are rotated and managers listen to workers' suggestions to ease the burden. Rather than being surveilled by the all-seeing eyes of those above them, co-owners are motivated by the will to work well for *their* business. Through sharing ownership and profits, workers gain greater control over their work, countering those crucial facets of deskilled workplaces. It is far from perfect and does not extend commoning out to those who eat the produce, but as a food commons in the meantime it shows how horticultural workers can be more fairly valued and rewarded.

Restoring skills ecosystems

In the previous chapter, I highlighted that horticulture's skills ecosystems are vulnerable because they rely on individual growers' viability. Solutions in the meantime work to share the risks and costs of knowledge exchange, investing public resources in skills development. Commoning ideals imply communities sharing responsibility and benefits, rather than state intervention and investment (Caffentzis and Federici 2014), but there may be a role for public investment in the meantime. Public funding already supports horticultural skills through training programmes: the one behind the IPM day described above was funded through Pillar II CAP subsidy, a dividend lost post-Brexit. Welsh Government has invested in Farming Connect, a public advisory system of the kind lost in England, meaning farmers and growers can access knowledge exchange covering technical production and business development, often free of charge. In contrast to this extension model, farmer to farmer knowledge exchange is a commons where knowledge is collectively developed and shared rather than flowing vertically from "the experts" (Rosset and Val 2020). But as apparent from knowledge sharing between smaller growers, this model is vulnerable within market-driven agri-food systems. As a commons in the meantime, groups of growers and related organisations have accessed public and charitable funding to invest in skills infrastructure, making peer learning more viable.

One training programme used EU structural funds to address some limitations of goodwill systems for learning agro-ecological horticulture. On-farm traineeships saw trainees work alongside an expert grower on their farm, learning by doing together and through hands-on experience. Trainee cohorts also came together for farm visits and group training sessions, and received accredited training in skills such as tractor driving. Growers were paid to deliver webinars and sessions sharing their expertise. The coordinating organisations supported on-farm trainers by developing a recommended curriculum for new growers' learning. A train-the-trainer element supported them to deliver more effective learning and to gain confidence as educators. Having third-party involvement gave trainees some protection from exploitation and informality which can plague internships (Ekers 2019; Weiler et al. 2016). It is fair to assume that being part of a network with the coordinating organsiations increased transparency, encouraging growers to provide quality learning opportunities and living conditions. Trainees achieved an accreditation which should help them secure horticultural jobs. Perhaps as important, they got to know experienced growers who they continue to look to for advice and inspiration. And maybe even more important, they formed a collective who have shared experiences of the training, getting to know each other through group training sessions, and in downtime when they cooked, drank and chatted around campfires. More than a year after the first cohort completed their traineeship, they were still in touch, using their WhatsApp group to share practical growing tips and more. Some are already training newer growers.

This training programme was not perfect; most significantly, it did not solve the problem of making sure traineeships are accessible and fair. Each farm had to find solutions to their challenge of affording to pay trainees as they worked. Younger trainees were eligible for a government programme which covered their pay for the period, but this was part of pandemic recovery funding no longer available. Many trainees were financially constrained during the traineeship; living on a low income in temporary accommodation for several months made it difficult to save for a housing deposit needed when moving for their next job.

Those involved in this example would like it to continue and be enhanced through financial support for the costs of employing trainees. There are good grounds for public investment, given that grower-trainers are feeding horticulture's skills ecosystem, contributing to the common good of growing good food. Paying expert growers to train novices rewards them for contributing their know-how and generates income for their business. These measures share risks and benefits instead of individualising them, enhancing the skills ecosystem by strengthening individual growers and by building a more robust infrastructure around skills development. But they do not address the barriers newly trained growers face to becoming food growers, such as access to rural housing and land where they can grow food. These problems reach beyond knowledge infrastructure and need appropriate solutions.

Knowledge commons in the meantime

Examples of horticultural enterprises collaborating, implementing collective models of food growing and learning demonstrate potential to move towards knowledge and food commons in the meantime. They do not achieve the multi-species commoning we might aspire to, but are developing alternatives and addressing challenges which thwart horticulture today. But it is difficult to shape more collective, shared systems of work and knowledge within the broader problems of the agri-food system which limit the power to thrive as a grower of good food. To go further, the sector needs to gain weight within decision-making which shapes food systems, so it becomes possible for food growers to achieve good, fair livelihoods. Building power with, through collectives advocating shared goals, is a vital strategy towards healthier horticulture.

Some steps to this seem straightforward, for example using public funds to pay growers to train future growers, in return for quality-assured traineeships, and reinstating collective investment in R&D to develop and exchange shared knowledge. Investing in other infrastructure around horticultural skills will help, for example further education colleges which deliver vital vocational learning. Longer strides going further are more difficult to conceive: how to correct deep-rooted misunderstandings of what horticultural work is like and how skilled it is? How to reshape food regimes so they offer quality jobs rather than precarious positions? We can start from commoning in the meantime, reversing tendencies to individualise skills and the blame for horticulture's workforce problems. A theme across this book has been that what growers do or not in relation to their workforce and knowledge ecosystem is constrained by macro forces connecting them into the pressures of global, commodity food systems. If we want diverse growers to keep growing food in the UK, they need assistance to ease this squeeze, through public support – financial and political – underpinning the vital contribution they make to society. As seen during 2022 (Sijmonsa 2022), when labour is unavailable, growers pause or reduce production, meaning more imports. New growers will be deterred from starting if a secure income is unlikely or start-up costs are prohibitive. For horticulture to thrive and expand requires political support, including recognition that food growing know-how is a collective good to be invested in as such. Horticultural knowledge is an essential common good, so research, training and education feeding the industry deserve public investment reflecting this and should be coordinated through collective oversight of the knowledge ecosystem. Stakeholders often note that the UK needs a strategy for how the country should feed itself; considering who will do the work and how they will access essential know-how should be central to this, to ensure resilience and share risks and benefits fairly.

Revaluing horticultural work requires a new contract for agricultural exceptionalism which flips the old form on its head. Growers should not be excused from controls on employment conditions because they cannot produce with them in place. Instead, controlling the quality of food production work should be regarded as exceptionally important because we all depend on what these workers produce. Everyone needs to eat, so the work which goes into producing it is a common good, which should be valued and rewarded by all.

References

Aberton, H. (2012). Material enactments of identities and learning in everyday community practices: Implications for pedagogy. *Pedagogy, Culture & Society*, *20*(1), 113–136. https://doi.org/10.1080/14681366.2012.649418

Adam & Eve (Director). (n.d.). *Waitrose & Partners Pick For Britain*. https://www.adamandeveddb.com/work/waitrose-and-partners-pick-for-britain

Adkins, F. (2020). The fruitless saga of the UK's 'Pick for Britain' scheme https://www.aljazeera.com/features/2020/11/19/pick-for-britain-a-rather-fruitless

Agyeman, J., & Giacalone, S. (2020). Introduction: The Immigrant-Food Nexus. In J. Agyeman & S. Giacalone (Eds.), *The Immigrant-Food Nexus: Borders, Labor, and Identity in North America* (pp. 1–15). The MIT Press. https://doi.org/10.7551/mitpress/11862.003.0003

AHDB. (2018). *Growing Champions in the next generation of pickers*. https://archive.ahdb.org.uk/growing-champions-in-the-next-generation-of-pickers

AHDB. (2020). *Horticulture Labour Barometer*. https://horticulture.ahdb.org.uk/coronavirus/horticulture-labour-barometer

AHDB. (n.d.). *Integrated pest management (IPM) hub*. https://ahdb.org.uk/integrated-pest-management-ipm-hub

Akram-Lodhi, A. H., & Kay, C. (2010). Surveying the agrarian question (part 1): Unearthing foundations, exploring diversity. *The Journal of Peasant Studies*, *37*(1), 177–202. https://doi.org/10.1080/03066150903498838

Aldous, D. E., Dixon, G. R., Darnell, R. L., & Pratley, J. E. (2014). Education and training futures in horticulture and horticultural science. In *Horticulture: Plants for People and Places, Volume 3: Social Horticulture* (pp. 1087–1115). Scopus. https://doi.org/10.1007/978-94-017-8560-0_32

Alkon, A. H. (2014). Food Justice and the Challenge to Neoliberalism. *Gastronomica*, *14*(2), 27–40. https://doi.org/10.1525/gfc.2014.14.2.27

Alkon, A. H., & Agyeman, J. (Eds.). (2011). *Cultivating food justice: Race, class, and sustainability*. MIT Press.

Allaire, G., & Daviron, B. (Eds.). (2019). *Ecology, capitalism and the new agricultural economy: The second great transformation*. Routledge, Taylor & Francis Group.

Allen, J. (2002). Living on Thin Abstractions: More Power/Economic Knowledge. *Environment and Planning A: Economy and Space*, *34*(3), 451–466. https://doi.org/10.1068/a3468

Allen, Patricia. (2016). Labor in the food system, from farm to table. *Journal of Agriculture, Food Systems, and Community Development*, *6*(2), 1. https://doi.org/10.5304/jafscd.2016.062.022

Anderson, B., Poeschel, F., & Ruhs, M. (2021). Rethinking labour migration: Covid-19, essential work, and systemic resilience. *Comparative Migration Studies*, *9*(1), 45–45. https://doi.org/10.1186/s40878-021-00252-2

Anderson, C. R., Maughan, C., & Pimbert, M. P. (2019). Transformative agroecology learning in Europe: Building consciousness, skills and collective capacity for food sovereignty. *Agriculture and Human Values*, *36*(3), 531–547. https://doi.org/10.1007/s10460-018-9894-0

Anderson, M. D. (2015). The role of knowledge in building food security resilience across food system domains. *Journal of Environmental Studies and Sciences*, *5*(4), 543–559. https://doi.org/10.1007/s13412-015-0311-3

Anderson, M. D. (2019). Scientific knowledge of food and agriculture in public institutions: Movement from public to private goods. In J. L. Vivero Pol, T. Ferrando, O. De Schutter, & U. Mattei (Eds.), *Routledge handbook of food as a commons*. Routledge.

Argüelles, L., & March, H. (2022). Weeds in action: Vegetal political ecology of unwanted plants. *Progress in Human Geography*, *46*(1), 44–66.

Armstrong, A. (1988). *Farmworkers: A social and economic history 1770-1980*. B. T. Batsford.

Atkins, L., Flint, K., & Oldfield, B. (2011). *Practical Matters: What young people think about vocational education in England*. City and Guilds Centre for Skills Development. http://www.skillsdevelopment.org/researchprojects/practical_matters.aspx#.VG4P6v3TTlI

Attewell, P. (1990). What is Skill? *Work and Occupations*, *17*(4), 422–448. https://doi.org/10.1177/0730888490017004003

Avallone, G. (2017). The land of informal intermediation. In A. Corrado, C. De Castro, & D. Perrotta (Eds.), *Migration and Agriculture* (pp. 217–230). Routledge.

Aznar-Sanchez, J., Belmonte-Ureña, L., & Tapia-Leon, J. (2014). The industrial agriculture: A "model for modernization" from Almeria? In J. Gertel & S. R. Sippel (Eds.), *Seasonal workers in mediterranean agriculture* (pp. 112–120). Earthscan.

Barber, P. (2014). Neoliberal market mystifications in the social costs of eating fresh – Phillippines-Canada comparisons. In J. Gertel & S. R. Sippel (Eds.), *Seasonal workers in mediterranean agriculture: The social costs of eating fresh* (pp. 95–100). Routledge.

Barndt, Deborah. (2008). *Tangled routes women, work, and globalization on the tomato trail* (2nd ed.). Rowman & Littlefield Pub.

Barnhill, A., & Doggett, T. (2018). Food ethics I: Food production and food justice. *Philosophy Compass*, *13*(3), e12479. https://doi.org/10.1111/phc3.12479

Batterham, J., & Levesley, T. (2011). *New directions: Young people's and parent's views of vocational education and careers guidance*. City & Guilds Centre for Skills Development. http://www.skillsdevelopment.org/pdf/New-directions-research-report.pdf

Binford, L. (2012). *Tomorrow we're all going to the harvest temporary foreign worker programs and neoliberal political economy* (1st ed.). University of Texas Press.

Boeckler, M., & Berndt, C. (2014). B/ordering the Mediterranean: Free trade, fresh fruits and fluid fixity. In *Seasonal Workers in Mediterranean Agriculture* (pp. 41–52). Routledge.

Böhm, S., Spierenburg, M., & Lang, T. (2020). Fruits of our labour: Work and organisation in the global food system. *Organization*, *27*(2), 195–212. https://doi.org/10.1177/1350508419888901

Bonanno, A., & Cavalcanti, J. S. B. (2014). Introduction. In A. Bonanno & J. S. B. Cavalcanti (Eds.), *Labour relations in globalized food* (pp. xiii–xlix). Emerald.

Bowler, I. (1985). *Some consequences of the industrialization of agriculture in the European Community* (pp. 75–98). The Industrialization of the Countryside. Norwich: Geobooks.

Braverman, H. (1998). *Labor and monopoly capital the degradation of work in the twentieth century* (25th anniversary ed.). Monthly Review Press.

Brice, J. (2014). Attending to grape vines: Perceptual practices, planty agencies and multiple temporalities in Australian viticulture. *Social & Cultural Geography*, *15*(8), 942–965. https://doi.org/10.1080/14649365.2014.883637

Brice, J. (2021). Vegetal labour and the measure of value: Reckoning time and producing worth in capitalist viticulture, In M. Ernwein, F. Ginn, & J. Palmer (Eds.), *The Work That Plants Do: Life, Labour and the Future of Vegetal Economies* (pp. 123–148). Transcript Verlag.

Broomfield, C., Nye, C., & Wells, R. (2022). Media framing of migrant labour in UK fruit and vegetable production: An analysis of reporting in UK farming and mainstream print press. *Journal of Rural Studies*, *95*, 423–437. https://doi.org/10.1016/j.jrurstud.2022.09.033

Brown, S., & Getz, C. (2008). Privatizing farm worker justice: Regulating labor through voluntary certification and labeling. *Geoforum*, *39*(3), 1184–1196. https://doi.org/10.1016/j.geoforum.2007.01.002

Brown, S., & Getz, C. (2011). Farmworker Food Insecurity and the Production of Hunger in California. In A. H. Alkon & J. Agyeman (Eds.), *Cultivating Food Justice: Race, Class and Sustainability* (pp. 121–146). MIT Press.

Brown, T. (2022). Skill ecosystems in the global South: Informality, inequality, and community setting. *Geoforum*, *132*, 10–19.

Bryson, J. (2017). Disciplinary Perspectives on Skill. In C. Warhurst, K. Mayhew, D. Finegold, & J. Buchanan (Eds.), *The Oxford Handbook of Skills and Training* (pp. 17–35). Oxford University Press.

Buchanan, J., Anderson, P., & Power, G. (2017). Skill ecosystems. In C. Warhurst, K. Mayhew, D. Finegold, & J. Buchanan (Eds.), *The Oxford Handbook of Skills and Training* (pp. 444–465). Oxford University Press.

Buchanan, K. (1948). Modern farming in the vale of Evesham. *Economic Geography*, *24*(4), 235–250.

Burch, D., & Lawrence, G. (2005). Supermarket own brands, supply chains and the transformation of the agri-food system. *International Journal of Sociology of Agriculture and Food*, *13*(1), 1–18.

Busch, L. (2019). The new autocracy in food and agriculture. In G. Allaire & B. Daviron (Eds.), *Ecology, Capitalism and the New Agricultural Economy* (pp. 95–110). Routledge.

Cadieux, K., & Slocum, R. (2016). Solidarity, space, and race: Toward geographies of agrifood justice. *Justice Spatiale | Spatial Justice*, *9*. https://www.jssj.org/article/solidarite-espace-et-race-vers-des-geographies-de-la-justice-alimentaire

Caffentzis, G., & Federici, S. (2014). Commons against and beyond capitalism. *Community Development Journal*, *49*, i92–i105.

CAHCR. (2018). *CAHCR Commodity Dashboard* [dataset]. https://app.powerbi.com/view?r=eyJrIjoiNDQyYTQ1MDgtY2QyNC00MmZmLTlhNWEtNzIzOGJmNDYyNmI2IiwidCI6ImM2OTdmYTEzLTY4OGQtNDEzNS1hMzgxLWFkNDEyMDY2ZWQ3YSIsImMiOjZ9

CAHCR. (2023). *Sowing seeds of change: Agriculture Labour Market Information Forecase to 2030*. CAHCR. https://cahrc-ccrha.ca/sites/default/files/2023-11/Executive%20report%20Final%20ENG%20%20%281%29.pdf

Calleja, E. J., Ilbery, B., & Mills, P. R. (2012). Agricultural change and the rise of the British strawberry industry, 1920–2009. *Journal of Rural Studies*, *28*(4), 603–611. https://doi.org/10.1016/j.jrurstud.2012.07.005

Calliste, J., Sivapragasam, S., McDonald, M., & Land In Our Names. (2021). *Rootz into growing*. Land In Our Names. https://static1.squarespace.com/static/5eece00ee6780d38b9fb012f/t/60ace24b8b87b828f51ec72e/1621942864883/FINAL+Rootz+into+Food+Growing+report+25.05.21.pdf

CALU. (2011). *Crop Walking: An introduction*. http://www.calu.bangor.ac.uk/Technical%20leaflets/080301%20CALU%20crop%20walking.pdf

Campbell, I. (2019). Harvest labour markets in Australia: Alleged labour shortages and employer demand for temporary migrant workers. *Journal of Australian Political Economy*, *84*, 46–88.

Campbell-Gibbons, H. (2011). Securing a sustainable British horticulture industry. *Nutrition Bulletin*, *36*(4), 443–448. https://doi.org/10.1111/j.1467-3010.2011.01934.x
Carlisle, L., Montenegro de Wit, M., DeLonge, M. S., Iles, A., Calo, A., Getz, C., Ory, J., Munden-Dixon, K., Galt, R., Melone, B., Knox, R., & Press, D. (2019). Transitioning to sustainable agriculture requires growing and sustaining an ecologically skilled workforce. *Frontiers in Sustainable Food Systems*, *3*. https://doi.org/10.3389/fsufs.2019.00096
Carnibella, F., & Wells, R. (2022). Framing of policy responses to migrant horticultural labour shortages during Covid-19 in the Italian print media. *Journal of Rural Studies*. https://doi.org/10.1016/j.jrurstud.2022.09.007
Carnoy, M., & Luschei, T. F. (2008). Skill acquisition in "high tech" export agriculture: A case study of lifelong learning in Peru's asparagus industry. *Journal of Education and Work*, *21*(1), 1–23. https://doi.org/10.1080/13639080801956982
Carolan, M. (2013). *The real cost of cheap food*. Earthscan.
Carolan, M. (2020). Automated agrifood futures: Robotics, labor and the distributive politics of digital agriculture. *The Journal of Peasant Studies*, *47*(1), 184–207. https://doi.org/10.1080/03066150.2019.1584189
Carolan, M. S. (2011). *Embodied food politics*. Ashgate.
Carr, C., & Gibson, C. (2016). Geographies of making: Rethinking materials and skills for volatile futures. *Progress in Human Geography*, *40*(3), 297–315. https://doi.org/10.1177/0309132515578775
Caruso, F.S. (2017). Unionism of migrant farm workers. The Sindicato Obreros del Campo (SOC) in Andalusia, Spain. In A. Corrado, C. de Castro, & D. Perrotta (Eds.), *Migration and agriculture* (pp. 277–292). Routledge.
Cavalcanti, J. S. B., & Bonanno, A. (2014). Conclusions: Labor between exploitation and resistance. In A. Bonanno & J. S. B. Cavalcanti (Eds.), *Research in Rural Sociology and Development* (Vol. 20, pp. 269–290). Emerald Group Publishing Limited. https://doi.org/10.1108/S1057-192220140000020011
Chamovitz, D. (2013). *What a plant knows: A field guide to the senses* (1st paperback ed). Scientific American/Farrar, Straus and Giroux.
Claeys, P., & Van Dyck, B. (2022). *ECVC Youth: Navigating Dreams & Precarity Working and Learning Conditions of Young Agricultural Workers, Interns and Volunteers Across Europe*. CAWR. https://www.eurovia.org/wp-content/uploads/2023/01/youth-report-EN-edit02.pdf
Clapp, J. (2022). *The rise of big food and agriculture: Corporate influence in the food system*(pp.45–66). Edward Elgar Publishing. https://doi.org/10.4337/9781800880269.00011
Clemens, M. A., Lewis, E. G., & Postel, H. M. (2018). Immigration Restrictions as Active Labor Market Policy: Evidence from the Mexican Bracero Exclusion. *The American Economic Review*, *108*(6), 1468–1487. https://doi.org/10.1257/aer.20170765
Clutterbuck, C. (2013). Worker Exploitation in Food Production and Service. In *Practical Ethics for Food Professionals* (pp. 153–169). John Wiley & Sons, Ltd.
Collyns, K. (2022). Labour and Wages: OGA survey. *The Organic Grower*, *58*, 28–31.
Constance, D. (2014). From fresh produce to poultry—Shifting labour regimes in the global agri-food system. In J. Gertel & S. Sippel (Eds.), *Seasonal workers in Mediterranean agriculture* (pp. 158–164). Routledge.
Cooke, G. (Ed.). (1981). *Agricultural Research 1931-1Ç81—History of the Agricultural Research Council and review of developments in agricultural science during the last 30 years*. Agricultural Research Council.
Corrado, A. (2017). Agrarian change and migrations in the Mediterranean from a food regime perspective. In A. Corrado, D. Perrotta, & C. de Castro (Eds.), *Migration and Agriculture Mobility and change in the Mediterranean area* (pp. 311–331). Routledge.

Corrado, A., de Castro, C., & Perrotta, D. (2017a). Introduction. Cheap food, cheap labour, high profits. Agriculture and mobility in the Mediterranean. In A. Corrado, C. de Castro, & D. Perrotta (Eds.), *Migration and Agriculture* (pp. 1–24). Routledge.

Corrado, A., de Castro, C., & Perrotta, D. (2017b). *Migration and Agriculture*. Routledge. https://doi.org/10.4324/9781315659558

Corrado, A., & Palumbo, L. (2022). Essential Farmworkers and the Pandemic Crisis: Migrant Labour Conditions, and Legal and Political Responses in Italy and Spain. In A. Triandafyllidou (Ed.), *Migration and Pandemics* (pp. 145–166). Springer International Publishing. https://doi.org/10.1007/978-3-030-81210-2_8

Coulter, S. (2018). Skill Formation, Immigration and European Integration: The Politics of the UK Growth Model. *New Political Economy*, *23*(2), 208–222. https://doi.org/10.1080/13563467.2017.1370446

Country, B., Wright, S., Suchet-Pearson, S., Lloyd, K., Burarrwanga, L., Ganambarr, R., Ganambarr-Stubbs, M., Ganambarr, B., Maymuru, D., & Sweeney, J. (2016). Co-becoming Bawaka: Towards a relational understanding of place/space. *Progress in Human Geography*, *40*(4), 455–475. https://doi.org/10.1177/0309132515589437

Cox, N., & Federici, S. (1975). *Counter-planning from the Kitchen: Wages for Housework: A Perspective on Capital and the Left*. Falling Wall Press.

Curry, N., Ingram, J., Kirwan, J., & Maye, D. (2012). Knowledge Networks for Sustainable Agriculture in England. *Outlook on Agriculture*, *41*(4), 243–248. https://doi.org/10.5367/oa.2012.0106

Curtain, R., M. Dornan, S. Howes, and H. Sherrell. (2018). Pacific Seasonal Workers: Learning from the Contrasting Temporary Migration Outcomes in Australian and New Zealand Horticulture. *Asia and the Pacific Policy Studies*, *5*(3), 462–480. https://doi.org/10.1002/app5.261.

Dalizel, P. (2017). Education and qualifications as skills. In D. Finegold, C. Warhurst, K. Mayhew, & J. Buchanan (Eds.), *The Oxford handbook of skills and training* (pp. 143–160). Oxford University Press.

Davies, A. (2022). Problems Continue in the Horticulture Sector: The Seasonal Workers Pilot Review 2019. *Industrial Law Journal (London)*., *51*(2), 494–499.

Davies, J. (2020). Corporate harm and embedded labour exploitation in agri-food supply networks. *European Journal of Criminology*, *17*(1), 70–85. https://doi.org/10.1177/1477370819874416

De Angelis, M. (2014). The commons. *Community Development Journal*, *49*, i68–i80.

de Castro, C., Reigada, A., & Gadea, E. (2020). The devaluation of female labour in fruit and vegetable packaging plants in Spanish Mediterranean agriculture. *Organization (London, England)*, *27*(2), 232–250.

Dedeurwaerdere, T., Frischmann, B., Hess, C., Lametti, D., Madison, M., Schweik, C., & Strandburg, K. (2014). *An Introduction to Knowledge Commons*. https://knowledge-commons.net/downloads/Knowledge%20Commons%20Description.pdf

DEFRA. (2019a). *Results from the 2018 Seasonal Labour in Horticulture End of Year Survey for England*. UK Government. https://assets.publishing.service.gov.uk/government/uploads/system/uploads/attachment_data/file/820589/labour-in-horticulture-survey-2018-final-25jul19.pdf

DEFRA. (2019b). *Results from the Seasonal Labour in Horticulture Survey 2018 for England*. https://assets.publishing.service.gov.uk/government/uploads/system/uploads/attachment_data/file/785863/Labour_in_Horticulture_Quarterly_Survey_2018_Final_14mar19.pdf

DEFRA. (2021a). *United Kingdom Food Security Report 2021: Theme 2: UK Food Supply Sources* [Official Statistics]. UK Government. https://www.gov.uk/government/statistics/united-kingdom-food-security-report-2021/united-kingdom-food-security-report-2021-theme-2-uk-food-supply-sources#united-kingdom-food-security-report-2021-theme2-indicator-2-1-9

DEFRA. (2021b). *Seasonal workers pilot review 2019* [Research and Analysis]. UK Government. https://www.gov.uk/government/publications/seasonal-workers-pilot-review/seasonal-workers-pilot-review-2019

DEFRA. (2022a). *Agriculture in the United Kingdom 2021*. UK Government. https://www.gov.uk/government/statistics/agriculture-in-the-united-kingdom-2021

DEFRA. (2022b). *Automation in horticulture review* [Independent Report]. UK Government.https://www.gov.uk/government/publications/defra-led-review-of-automation-in-horticulture/automation-in-horticulture-review

DEFRA. (2022c). *Consultation outcome Agricultural and Horticultural Development Board (AHDB) reform*. UK Government. https://www.gov.uk/government/consultations/agricultural-and-horticultural-development-board-ahdb-reform/outcome/executive-summary-analysis-of-responses-to-the-consultation-and-what-we-will-do-next

DEFRA. (2022d). *Horticulture Statistics 2021*. https://assets.publishing.service.gov.uk/government/uploads/system/uploads/attachment_data/file/1076378/hort-report-19may22.odt

DEFRA. (2022e). *Numbers of holdings and agricultural activity by farm type at 1 June each year in England*. UK Government. https://assets.publishing.service.gov.uk/government/uploads/system/uploads/attachment_data/file/1084964/structure-england-june21-farmtype-23jun22.ods

DEFRA. (2023). *Independent Review into Labour Shortages in the Food Supply Chain*. DEFRA. https://assets.publishing.service.gov.uk/government/uploads/system/uploads/attachment_data/file/1166600/Independent_Review_into_Labour_Shortages_in_the_Food_Supply_Chain_June_2023.pdf

Deutsch, T. (2012). 61 Labor Histories of Food. In J. M. Pilcher (Ed.), *The Oxford Handbook of Food History* (p. 0). Oxford University Press. https://doi.org/10.1093/oxfordhb/9780199729937.013.0004

Devlin, S. (2016). *Agricultural labour in the UK*. Food Research Collaboration. https://foodresearch.org.uk/publications/agricultural-labour-in-the-uk/

Directive (EU) 2019/1152 of the European Parliament and of the Council of 20 June 2019 on transparent and predictable working conditions in the European Union, (2019). https://eur-lex.europa.eu/legal-content/EN/TXT/?uri=CELEX:32019L1152

Dixon, G. R. (2016). *Making 21st century European horticultural education fit-for-purpose* (Vol. 1126). Scopus. https://doi.org/10.17660/ActaHortic.2016.1126.12

Dugan, E. (2022). Indonesians wait for UK farm jobs after paying deposits of up to £2,500. *The Guardian*. https://www.theguardian.com/uk-news/2022/sep/25/indonesians-wait-on-uk-farm-jobs-after-paying-deposits-of-up-to-2500

Dumont, A. M., & Baret, P. V. (2017). Why working conditions are a key issue of sustainability in agriculture? A comparison between agroecological, organic and conventional vegetable systems. *Journal of Rural Studies*, *56*, 53–64. https://doi.org/10.1016/j.jrurstud.2017.07.007

Dun, O., & Klocker, N. (2017). The Migration of Horticultural Knowledge: Pacific Island seasonal workers in rural Australiaa missed opportunity? *Australian Geographer*, *48*(1), 27–36. https://doi.org/10.1080/00049182.2017.1272528

EFFP. (2011). *Getting to the heart of horticulture: Opportunities and challenges for the horticulture and potato sectors in the West Midlands*. https://www.nfuonline.com/final_report_-_heart_of_horticulture_2011_1-sflb/

EFRA (2017). Feeding the nation: labour constraints https://publications.parliament.uk/pa/cm201617/cmselect/cmenvfru/1009/100902.htm

Ekers, M. (2019). The curious case of ecological farm interns: On the populism and political economy of agro-ecological farm work. *The Journal of Peasant Studies*, *46*(1), 21–43. https://doi.org/10.1080/03066150.2018.1512487

Ekers, M., Levkoe, C. Z., Walker, S., & Dale, B. (2016). Will work for food: Agricultural interns, apprentices, volunteers, and the agrarian question. *Agriculture and Human Values*, *33*(3), 705–720. https://doi.org/10.1007/s10460-015-9660-5

Ericksen, P. J. (2008). Conceptualizing food systems for global environmental change research. *Global Environmental Change*, *18*(1), 234–245. https://doi.org/10.1016/j.gloenvcha.2007.09.002

Ernwein, M., Palmer, J., & Ginn, F. (2021). Introduction: The work that plants do. In M. Ernwein, F. Ginn, & J. Palmer (Eds.), *The work that plants do* (pp. 15–31). Transcript Verlag.

EUROFOUND. (2006). *Industrial relations in agriculture (Infosheet)*. https://www.eurofound.europa.eu/en/publications/2012/industrial-relations-agriculture-infosheet

European Commission. (n.d.). *Factsheet – a greener and fairer CAP*. https://agriculture.ec.europa.eu/document/download/89b607ec-8a43-4073-bafd-f2493da7699e_en?filename=factsheet-newcap-environment-fairness_en.pdf

European Parliament. (2019). *The EU fruit and vegetable sector Main features, challenges and prospects*. https://www.europarl.europa.eu/RegData/etudes/BRIE/2019/635563/EPRS_BRI(2019)635563_EN.pdf

European Parliament. (2021). *Migrant seasonal workers in the European agricultural sector*. https://www.europarl.europa.eu/RegData/etudes/BRIE/2021/689347/EPRS_BRI(2021)689347_EN.pdf

Eurostat. (2020). *People on the move STATISTICS ON MOBILITY IN EUROPE* [dataset]. https://ec.europa.eu/eurostat/cache/digpub/eumove/index.html?lang=en

Eurostat. (2022). *Key figures on the European food chain – 2022 edition*. https://ec.europa.eu/eurostat/documents/15216629/15559935/KS-FK-22-001-EN-N.pdf/1cb9d295-6868-70e3-0319-4725040cfdb8?version=3.0&t=1670599965263

Eurostat. (2023). *Migration and asylum in Europe – 2023 edition* [Dataset]. https://ec.europa.eu/eurostat/web/interactive-publications/migration-2023

FAO. (2020a). *Fruit and vegetables – your dietary essentials*. FAO. https://doi.org/10.4060/cb2395en

FAO. (2020b). *Migrant workers and the COVID-19 pandemic*. FAO. https://www.fao.org/documents/card/en/c/ca8559en

FAO. (2021). The International Year of Fruits and Vegetables, 2021. https://www.fao.org/fruits-vegetables-2021/en

FAO. (2022). *World Food and Agriculture – Statistical Yearbook 2022*. FAO. https://www.fao.org/documents/card/en/c/cc2211en

Federici, S. (2011). Women, land struggles, and the reconstruction of the commons. *Working USA*, *14*(1), 41–56. https://doi.org/10.1111/j.1743-4580.2010.00319.x

Federici, S., & Linebaugh, P. (2019). *Re-enchanting the world: Feminism and the politics of the commons*. PM Press.

Feinberg, R. M. (2020). The new contadini: Transformative labor in Italian vineyards. *Agriculture and Human Values*, *38*(1), 15–28. https://doi.org/10.1007/s10460-020-10125-6

Findlay, A., & McCollum, D. (2013). Recruitment and employment regimes: Migrant labour channels in the UK's rural agribusiness sector, from accession to recession. *Journal of Rural Studies*, *30*, 10–19.

Finegold, D. (1999). Creating self-sustaining, high-skill ecosystems. *Oxford Review of Economic Policy*, *15*(1), 60–81. https://doi.org/10.1093/oxrep/15.1.60

Fischer, E. & Benson, P. (2006). *Brocolli and desire: Global connections and Maya struggles in postwar Guatemala*. Stadford University Press.

Fisher, T., & Botticello, J. (2018). Machine-made lace, the spaces of skilled practices and the paradoxes of contemporary craft production. *Cultural Geographies*, *25*(1), 49–69. https://doi.org/10.1177/1474474016680106

Fitzgerald, D. K. (2003). *Every farm a factory: The industrial ideal in American agriculture*. Yale University Press. https://doi.org/10.12987/9780300133417

FLEX. (2021). *Assessment of the Risks of Human Trafficking for Forced Labour on the UK Seasonal Workers Pilot*. https://labourexploitation.org/publications/assessment-risks-human-trafficking-forced-labour-uk-seasonal-workers-pilot#overlay-

context=publications/assessment-risks-human-trafficking-forced-labour-uk-seasonal-workers-pilot
FLEX (2022). Flex, written evidence to 'Labour shortages in the food and farming sector' 2022, all written evidence available at https://committees.parliament.uk/work/1497/labour-shortages-in-the-food-and-farming-sector/publications/
Fogg, G. (1963). *The Growth of Plants*. Pelican an imprint of Penguin Books.
Folke, C., Carpenter, S. R., Walker, B., Scheffer, M., Chapin, T., & Rockstrom, J. (2010). Resilience Thinking: Integrating Resilience, Adaptability and Transformability. *Ecology and Society*, *15*(4), Art. 20. https://doi.org/10.5751/ES-03610-150420
Fortier, J.-M. (2014). *The market gardener: A successful grower's handbook for small-scale organic farming* (S. Irving, Trans.). New Society Publishers.
Fox Miller, C. (2017). The contemporary geographies of craft-based manufacturing. *Geography Compass*, *11*(4), np-n/a. https://doi.org/10.1111/gec3.12311
Freidberg, S. (2009). *Fresh a perishable history*. Belknap Press of Harvard University Press. https://doi.org/10.4159/9780674053854
Freidberg, S. (2003). Cleaning up down South: Supermarkets, ethical trade and African horticulture. *Social & Cultural Geography*, 4(1), 27–43. https://doi.org/10.1080/1464936032000049298
Friedland, W. (1994). The Global Fresh Fruit and Vegetable System: An industrial organisational analysis. In McMichael, P. (Ed.), *The global restructuring of agro-food systems* (p. 173). Cornell University Press.
Friedland, W. H., Barton, A. E., & Thomas, R. J. (1981). *Manufacturing green gold: Capital, labor, and technology in the lettuce industry*. Cambridge University Press.
Friedmann, H. (2006). From Colonialism to Green Capitalism: Social Movements and Emergence of Food Regimes. In *Research in Rural Sociology and Development* (Vol. 11, pp. 227–264). Emerald (MCB UP). https://doi.org/10.1016/S1057-1922(05)11009-9
Friedmann, H., & McMichael, P. (1989). Agriculture and the State System: The Rise and Decline of National Agricultures, 1870 to the Present',XXIX (2), 93-117. *Sociologia Ruralis*, *14*(2), 93–117.
Fuglie, K. O., & Toole, A. A. (2014). The Evolving Institutional Structure of Public and Private Agricultural Research. *American Journal of Agricultural Economics*, *96*(3), 862–883. https://doi.org/10.1093/ajae/aat107
Gadea, E., Pedreno, A., & De Castro, C. (2017). Producing and mobilizing vulnerable workers. In A. Corrado, C. De Castro, & D. Perrotta (Eds.), *Migration and Agriculture* (pp. 79–94). Routledge.
Gagliano, M. (2013). Seeing Green: The Re-discovery of Plants and Nature's Wisdom. *Societies*, *3*(1), 147–157. https://doi.org/10.3390/soc3010147
Gagliano, M., Renton, M., Depczynski, M., & Mancuso, S. (2014). Experience teaches plants to learn faster and forget slower in environments where it matters. *Oecologia*, *175*(1), 63–72. https://doi.org/10.1007/s00442-013-2873-7
Galt, R. E. (2013). The Moral Economy Is a Double-edged Sword: Explaining Farmers' Earnings and Self-exploitation in Community-Supported Agriculture. *Economic Geography*, *89*(4), 341–365. https://doi.org/10.1111/ecge.12015
García-López, G. A., Lang, U., & Singh, N. (2021). Commons, Commoning and Co-Becoming: Nurturing Life-in-Common and Post-Capitalist Futures (An Introduction to the Theme Issue). *Environment and Planning E: Nature and Space*, *4*(4), 1199–1216. https://doi.org/10.1177/25148486211051081
Garnett, P., Doherty, B., & Heron, T. (2020). Vulnerability of the United Kingdom's food supply chains exposed by COVID-19. *Nature Food*, *1*(6), 315–318. https://doi.org/10.1038/s43016-020-0097-7
Garth, Hanna & Reese, Ashanté M. (2020). *Black Food Matters*. U. of Minnesota Press.
Gerbeau, Y. M., & Avallone, G. (2016). Producing Cheap Food and Labour: Migrations and Agriculture in the Capitalistic World-Ecology. *Social Change Review*, *14*(2), 121–148. https://doi.org/10.1515/scr-2016-0025

Gertel, J., & Sippel, S. R. (2014). Super/markets: Beyond buyer-drivenness in southern France. In Jörg Gertel, & Sarah Ruth Sippel (Eds.), *Seasonal Workers in Mediterranean Agriculture* (pp. 67–78). Routledge.

Gertler, M. S. (2003). Tacit knowledge and the economic geography of context, or the undefinable tacitness of being (there). *Journal of Economic Geography*, *3*(1), 75–99.

Getz, C., Brown, S., & Shreck, A. (2008). Class Politics and Agricultural Exceptionalism in California's Organic Agriculture Movement. *Politics & Society*, *36*(4), 478–507. https://doi.org/10.1177/0032329208324709

Gherardi, S. (2006). *Organizational knowledge the texture of workplace learning*. Malden, MA: Blackwell Pub.

Gibson, J., & Bailey, R.-L. (2021). Seasonal Labor Mobility in the Pacific: Past Impacts, Future Prospects. *Asian Development Review*, *38*(1), 1–31. https://doi.org/10.1162/adev_a_00156

Gibson-Graham, J. K. (2006). *A postcapitalist politics*. University of Minnesota Press.

Gieser, T. (2008). Embodiment, emotion and empathy: A phenomenological approach to apprenticeship learning. *Anthropological Theory*, *8*(3), 299–318. https://doi.org/10.1177/1463499608093816

Gieser, T. (2014). Enskillment inhibited: "industrial gardening" in Britain. *Journal of the Royal Anthropological Institute*, *20*(1), 131–149. https://doi.org/10.1111/1467-9655.12083

Gill, A. H. (1988). The Horticultural Development Council in Great Britain. *Acta Horticulturae*, *223*, 365–371. https://doi.org/10.17660/ActaHortic.1988.223.58

Gilmore, R. (1999). Globalisation and US prison growth: From military Keynesianism to post-Keynesian militarism. *Race & Class.*, *40*(2/3).

Ginn, F. (2012). Dig for Victory! New histories of wartime gardening in Britain. *Journal of Historical Geography*, *38*(3), 294–305. https://doi.org/10.1016/j.jhg.2012.02.001

Glenna, L. (2020). The value of public agricultural and food knowledge during pandemics. *Agriculture and Human Values*, *37*(3), 607–608. https://doi.org/10.1007/s10460-020-10091-z

Gonçalves, M. (2023, March 16). Labour agencies to face supermarket scrutiny over exploitation claims. *The Grocer*. https://www.thegrocer.co.uk/supply-chain/labour-agencies-to-face-supermarket-scrutiny-over-exploitation-claims/677331.article

Good work for good food. (2021). https://the.organise.network/campaigns/teamup-good-work-for-good-food

Gottlieb, R., & Joshi, A. (2013). *Food justice* (1st MIT Press paperback ed). The MIT Press.

Graddy-Lovelace, G. (2020). Plants: Crop diversity pre-breeding technologies as agrarian care co-opted? *Area*, *52*, 235–243. https://doi.org/10.1111/area.12499

Gray, M. (2014). *Labor and the locavore: The making of a comprehensive food ethic*. University of California Press. https://doi.org/10.1525/9780520957060

Greco, E. (2020). Global value relations and local labour control regimes in rice farming in Uganda and Tanzania. *Organization*, *27*(2), 213–231. https://doi.org/10.1177/1350508419888902

Green, F. (2011). *What is Skill? An Inter-Disciplinary Synthesis*. Centre for Learning and Life Chances in Knowledge Economies and Societies. http://www.llakes.org

Growers Better Levy Group. (n.d.). *A Vision for the Future of UK Horticulture*. https://britishgrowers.org/wp-content/uploads/2021/03/GBLG-Vision-for-the-Future-of-UK-Horticulture-1.pdf

Guthman, J. (2014). *Agrarian dreams: The paradox of organic farming in California* (2nd ed.). University of California Press. https://doi.org/10.1525/9780520959132

Guthman, J. (2017). Willing (White) Workers on Organic Farms? Reflections on Volunteer Farm Labor and the Politics of Precarity. *Gastronomica*, *17*(1), 15. https://doi.org/10.1525/gfc.2017.17.1.15

Hahamovitch, C. (2003). Creating perfect immigrants: Guestworkers of the world in historical perspective. *Labor History*, *44*(1), 69–94. https://doi.org/10.1080/0023656032000057010

Hall, M. (2011). *Plants as persons: A philosophical botany*. State University of New York Press.

Head, L., & Atchison, J. (2009). Cultural ecology: Emerging human-plant geographies. *Progress in Human Geography*, *33*(2), 236–245. https://doi.org/10.1177/0309132508094075

Healey, M. J., Ilbery, B. W., Institute of British Geographers, & Institute of British Geographers (Eds.). (1985). *The Industrialization of the countryside*. Geo Books.

Hedberg, C. (2014). Restructuring global labor markets: Recruitment agencies and work relations in the wild berry commodity chain. In A. Bonanno & J. S. B. Cavalcanti (Eds.), *Research in Rural Sociology and Development* (Vol. 20, pp. 33–55). Emerald Group Publishing Limited. https://doi.org/10.1108/S1057-192220140000020000

Helfrich, S., & Haas, J. (2009). *The Commons: A New Narrative for Our Times*. Heinrich Boll Stiftung. http://us.boell.org/sites/default/files/downloads/CommonsBook_Helfrich_-_Haas-neu.pdf

Hellio, E. (2014). 'We don't have women in boxes': Channelling seasonal mobility of female farmworkers between Morocco and Andalusia. In Jörg Gertel, & Sarah Ruth Sippel (Eds.), *Seasonal Workers in Mediterranean Agriculture* (pp. 159–182). Routledge.

Hernández Romero, M. A. (2012). Nothing to Learn? Labor Learning in California's Farmwork. *Anthropology of Work Review*, *33*(2), 73–88. https://doi.org/10.1111/j.1548-1417.2012.01081.x

Hingley, M. K. (2005). Power imbalanced relationships: Cases from UK fresh food supply. *International Journal of Retail & Distribution Management*, *33*(8/9), 551–569. https://doi.org/10.1108/09590550510608368

Hitchings, R. (2012). People can talk about their practices. *Area (London 1969)*, *44*(1), 61–67. https://doi.org/10.1111/j.1475-4762.2011.01060.x

Holmes, S. M. (2013). *Fresh fruit, broken bodies migrant farmworkers in the United States*. University of California Press. https://doi.org/10.1525/9780520954793

Holt-Gimenez, E., & van Lammeren, I. (2020). Can food as a commons advance food sovereignty? In Jose Luis Vivero-Pol, Tomaso Ferrando, Olivier De Schutter, & Ugo Mattei (Eds.), *The Routledge Handbook of Food as Commons* (pp. 313–328). Routledge.

Hornborg, A. (2018). Relationism as revelation or prescription? Some thoughts on how Ingold's implicit critique of modernity could be harnessed to political ecology. *Interdisciplinary Science Reviews*, 1–11. https://doi.org/10.1080/03080188.2018.1524685

Horticultural Sector Committee. (2023). *Sowing the seeds: A blooming English horticultural sector*. House of Lords. https://publications.parliament.uk/pa/ld5803/ldselect/ldhortcom/268/268.pdf

Horticulture Matters. (2014). *Horticulture Matters*. RHS. https://www.rhs.org.uk/education-learning/pdf/Training/1016-rhs-hort-careers-brochure-v8.pdf

Horticulture New Zealand. (2023). *Growing Together 2035 Aotearoa Horticulture Action Plan – Strategy*. Horticulture New Zealand. https://www.hortnz.co.nz/assets/About-Us/Aotearoa-Horticulture-Action-Plan/HOR-4619_Horticulture-Action-Plan-FA_web.pdf

Ilbery, B., & Bowler, I. (1994). International competition in agriculture and government intervention: The case of horticulture in England. *Geography*, *79*(4), 361–366.

Ilbery, B., & Bowler, I. R. (1995). Britain's Uncertain Future in the International Horticultural Market. *Outlook on Agriculture*, *24*(1), 49–55. https://doi.org/10.1177/003072709502400110

Ilbery, B. W. (1985a). Factors affecting the structure of horticulture in the Vale of Evesham, U.K.: A behavioural interpretation. *Journal of Rural Studies*, *1*(2), 121–133. https://doi.org/10.1016/0743-0167(85)90065-8

Ilbery, B. W. (1985b). Horticultural Decline in the Vale of Evesham, U.K., 1950–1980. *Journal of Rural Studies*, *1*(2), 109–120. https://doi.org/10.1016/0743-0167(85)90064-6

Independent Chief Inspector of Borders and Immigration. (2022). *An inspection of the immigration system as it relates to the agricultural sector*. https://assets.publishing.service.gov.uk/government/uploads/system/uploads/attachment_data/file/1125411/An_inspection_of_the_immigration_system_as_it_relates_to_the_agricultural_sector_May_to_August_2022.pdf

Independent Review into Labour. (2023). *Independent Review into Labour Shortages in the Food Supply Chain*. DEFRA. https://assets.publishing.service.gov.uk/government/uploads/system/uploads/attachment_data/file/1166600/Independent_Review_into_Labour_Shortages_in_the_Food_Supply_Chain_June_2023.pdf

Independent Review into Labour Shortages in the Food Supply Chain. (2023). DEFRA. https://assets.publishing.service.gov.uk/government/uploads/system/uploads/attachment_data/file/1166600/Independent_Review_into_Labour_Shortages_in_the_Food_Supply_Chain_June_2023.pdf

Ingold, T. (1997). Eight themes in the anthropology of technology. *Social Analysis: The International Journal of Social and Cultural Practice*, *41*(1), 106–138.

Ingold, T. (2000). *The perception of the environment: Essays on livelihood, dwelling & skill*. Routledge.

Ingold, T. (2011). *Being alive: Essays on movement, knowledge and description*. Routledge.

Ingold, T. (2016). From science to art and back again: The pendulum of an anthropologist. *Anuac*, *5*, 5–23. https://doi.org/10.7340/ANUAC2239-625X-2237

Ingram, J. (2008). Are farmers in England equipped to meet the knowledge challenge of sustainable soil management? An analysis of farmer and advisor views. *Journal of Environmental Management*, *86*(1), 214–228. https://doi.org/10.1016/j.jenvman.2006.12.036

Iskander, N. N. (2021). *Does skill make us human? Migrant workers in 21st-century Qatar and beyond*. Princeton University Press.

Jack, L., & Hammans, H. (2022). *Unpicking food prices: Where does your food pound go, and why do farmers get so little?* Sustain. https://www.sustainweb.org/reports/dec22-unpicking-food-prices/

Keep, E., & Mayhew, K. (2010). Moving beyond skills as a social and economic panacea. *Work, Employment and Society*, *24*(3), 565–577.

Kimmerer, R. W. (2013). *Braiding sweetgrass: Indigenous wisdom, scientific knowledge and the teachings of plants* (First paperback edition). Milkweed Editions.

Klassen, S., Medland, L., Nichol, P., & Pitt, H. (2023). Pathways for advancing good work in food systems: Reflecting on the international Good Work for Good Food Forum. *Journal of Agriculture, Food Systems, and Community Development*, 1–17. https://doi.org/10.5304/jafscd.2023.122.004

Klerkx, L., Jakku, E., & Labarthe, P. (2019). A review of social science on digital agriculture, smart farming and agriculture 4.0: New contributions and a future research agenda. *NJAS - Wageningen Journal of Life Sciences*, *90–91*, 100315. https://doi.org/10.1016/j.njas.2019.100315

Klocker, N., Dun, O., Head, L., & Gopal, A. (2019). Exploring migrants' knowledge and skill in seasonal farm work: More than labouring bodies. *Agriculture and Human Values*. https://doi.org/10.1007/s10460-019-10001-y

Krzywoszynska, A. (2016). What Farmers Know: Experiential Knowledge and Care in Vine Growing: Experiential knowledge and care in vine growing. *Sociologia Ruralis*, *56*(2), 289–310. https://doi.org/10.1111/soru.12084

Krzywoszynska, A. (2019). Making knowledge and meaning in communities of practice: What role may science play? The case of sustainable soil management in England. *Soil Use and Management*, *35*(1), 160–168. https://doi.org/10.1111/sum.12487

Kuns, B., Börjeson, L., Fischer, K., Hedberg, C., Olofsson, I., Ovaska, U., Refslund, B., Fredrik Rye, J., & Vihinen, H. (2023). From panic to business as usual: What coronavirus has revealed about migrant labour, agri-food systems and industrial relations in the Nordic countries. *Sociologia Ruralis*, *63*(4), 907–927. https://doi.org/10.1111/soru.12443

Lam, A., & Marsden, D. (2017). Employment systems, skills, and knowledge. In C. Warhurst, K. Mayhew, D. Finegold, & J. Buchanan (Eds.), *The Oxford Handbook of Skills and Training* (pp. 466–487). Oxford University Press.

Landworkers Alliance. (2017). *Hands on the Land New Entrants in Agriculture*. https://landworkersalliance.org.uk/wp-content/uploads/2019/07/LWA-New-Entrant-Policy-Paper-August-2017.pdf

Landworkers Alliance. (2024). *Traineeships update*. https://landworkersalliance.org.uk/wp-content/uploads/2018/10/Traineeships-update-March-2024.pdf

Lang, T. (2020). *Feeding Britain: Our food problems and how to fix them*. Penguin Books.

Lauder, H., Brown, P., & Ashton, D. (2017). Theorizing skill formation in the global economy. In J. Buchanan, D. Finegold, K. Mayhew, & C. Warhurst (Eds.), *The Oxford handbook of skills and training* (Vol. 1). Oxford University Press. https://doi.org/10.1093/oxfordhb/9780199655366.013.19

Lave, J., & Wenger, E. (1991). *Situated learning: Legitimate peripheral participation*. Cambridge University Press.

Lawrence, A. M. (2022). Listening to plants: Conversations between critical plant studies and vegetal geography. *Progress in Human Geography*, *46*(2), 629–651. https://doi.org/10.1177/03091325211062167

Lee, R. (2014). European food governance. In T. Marsden & A. Morley (Eds.), *Sustainable food systems: Building a new paradigm* (pp. 78–99). Routledge.

Legun, K., Burch, K. A., & Klerkx, L. (2023). Can a robot be an expert? The social meaning of skill and its expression through the prospect of autonomous AgTech. *Agriculture and Human Values*, *40*(2), 501–517. https://doi.org/10.1007/s10460-022-10388-1

Leitch Review. (2006). *Final Report Prosperity for all in the global economy—World class skills*. HMSO. https://assets.publishing.service.gov.uk/government/uploads/system/uploads/attachment_data/file/243599/0118404865.pdf

Leitheiser, S., Trell, E.-M., Horlings, I., & Franklin, A. (2022). Toward the commoning of governance. *Environment and Planning C: Politics and Space*, *40*(3), 744–762. https://doi.org/10.1177/23996544211033992

Lever, J., & Milbourne, P. (2017). The Structural Invisibility of Outsiders: The Role of Migrant Labour in the Meat-Processing Industry. *Sociology (Oxford)*, *51*(2), 306–322. https://doi.org/10.1177/0038038515616354

Liu-Farrer, G., Yeoh, B. S., & Baas, M. (2021). Social construction of skill: An analytical approach toward the question of skill in cross-border labour mobilities. *Journal of Ethnic and Migration Studies*, *47*(10), 2237–2251. https://doi.org/10.1080/1369183X.2020.1731983

Lowe, P., Phillipson, J., Proctor, A., & Gkartzios, M. (2019). Expertise in rural development: A conceptual and empirical analysis. *World Development*, *116*, 28–37. https://doi.org/10.1016/j.worlddev.2018.12.005

Mabey, R. (2011). *Weeds: In defense of nature's most unloved plants* (1st U.S. ed.). HarperCollins.

MacAuley, L., & Niewolny, K. (2016). Situating On-farm Apprenticeships within the Alternative Agrifood Movement: Labor and Social Justice Implications. *Journal of Agriculture, Food Systems, and Community Development*, *6*(2), 195–223. https://doi.org/10.5304/jafscd.2016.062.024

Magnan, A. (2012). 370 Food Regimes. In J. M. Pilcher (Ed.), *The Oxford Handbook of Food History* (p. 0). Oxford University Press. https://doi.org/10.1093/oxfordhb/9780199729937.013.0021

Mancuso, S., Viola, A., Benham, J., & Pollan, M. (2015). *Brilliant green: The surprising history and science of plant intelligence*. Island Press.

Marder, M. (2013). *Plant-Thinking A Philosophy of Vegetal Life*. Columbia University Press.
Martin, J. (2000). *The development of modern agriculture: British farming since 1931*. St. Martin's Press.
Martin, P. L. (2016). *Migrant workers in commercial agriculture*. International Labour Office. https://www.ilo.org/wcmsp5/groups/public/---ed_protect/---protrav/---migrant/documents/publication/wcms_538710.pdf
Martin, P. L. (2021). *The prosperity paradox fewer and more vulnerable farm workers* (1st ed.). Oxford University Press.
Maye, D. (2018). Examining Innovation for Sustainability from the Bottom Up: An Analysis of the Permaculture Community in England. *Sociologia Ruralis*, *58*(2), 331–350. https://doi.org/10.1111/soru.12141
Mazé, A., Calabuig Domenech, A., & Goldringer, I. (2021). Commoning the seeds: Alternative models of collective action and open innovation within French peasant seed groups for recreating local knowledge commons. *Agriculture and Human Values*, *38*(2), 541–559. https://doi.org/10.1007/s10460-020-10172-z
McAndrew, C., Fisher, O., McCallister, C., & Jaccarini, C. (2023). *Debt, Migration and Exploitation*. Landworkers Alliance. https://staging.landworkersalliance.org.uk/wp-content/uploads/2018/10/LWA-Debt-Migration-and-Exploitation-2023.pdf?_gl=1*1hi86b5*_ga*MjEzMTg4NDM0MC4xNjk3MTk3ODEw*_ga_ZDW3NFY5G0*MTcwMDIzNzcwMS40LjAuMTcwMDIzNzcwMS4wLjAuMA.*_ga_HGZDP7MC4L*MT|cwMDIzNzcwMS40LjAuMTcwMDIzNzcwMS4wLjAuMA.&_ga=2.205331344.1834874359.1700237702-2131884340.1697197810
McGuinness, T., & Garton Grimwood, G. (2017). *Migrant workers in agriculture, Commons Library Briefing Paper CBP-7987*. Parliament UK. https://researchbriefings.files.parliament.uk/documents/CBP-7987/CBP-7987.pdf
McLaughlin, J., & Weiler, A. M. (2017). Migrant agricultural workers in local and global contexts: Toward a better life? *Journal of Agrarian Change*, *17*(3), 630–638. https://doi.org/10.1111/joac.12199
McMichael, P. (1997). Rethinking globalization: The agrarian question revisited. *Review of International Political Economy*, *4*(4), 630–662. https://doi.org/10.1080/096722997708565786
Medland, L. (2021). 'There is no time': Agri-food internal migrant workers in Morocco's tomato industry. *Journal of Rural Studies*, *88*, 482–490. https://doi.org/10.1016/j.jrurstud.2021.04.015
Mellino, E., & Pangeni, R. (2023, March 27). *'They treat you like an animal': How British farms run on exploitation*. The Bureau of Investigative Journalism. https://www.thebureauinvestigates.com/stories/2023-03-27/they-treat-you-like-an-animal-how-british-farms-run-on-exploitation
Mellino, E., Pattison, P., & Pangeni, R. (2022, May 27). Migrant fruit pickers charged thousands in illegal fees to work on UK farms, investigation shows. *The Guardian*. https://www.theguardian.com/global-development/2022/may/27/migrant-fruit-pickers-charged-thousands-in-fees-to-work-on-uk-farms-investigation-shows?CMP=share_btn_tw
Melossi, E. (2021). 'Ghetto tomatoes' and 'taxi drivers': The exploitation and control of Sub-Saharan African migrant tomato pickers in Puglia, Southern Italy. *Journal of Rural Studies*, *88*, 491–499. https://doi.org/10.1016/j.jrurstud.2021.04.009
Menary, J., Collier, R., & Seers, K. (2019). Innovation in the UK fresh produce sector: Identifying systemic problems and the move towards systemic facilitation. *Agricultural Systems*, *176*, 102675. https://doi.org/10.1016/j.agsy.2019.102675
Mésini, B. (2014). The transnational recruitment of temporary Latino workers in European agriculture. In Jörg Gertel, & Sarah Ruth Sippel (Eds.), *Seasonal Workers in Mediterranean Agriculture* (pp. 89–100). Routledge.

Mieres, F., & McGrath, S. (2021). Ripe to be heard: Worker voice in the Fair Food Program. *International Labour Review*, *160*(4), 631–647. https://doi.org/10.1111/ilr.12204

Migration Advisory Committee. (2020). *Review of the Shortage Occupation List, 2020 September 2020*. https://assets.publishing.service.gov.uk/government/uploads/system/uploads/attachment_data/file/927352/SOL_2020_Report_Final.pdf

Milbourne, P., & Coulson, H. (2021). Migrant labour in the UK's post-Brexit agri-food system: Ambiguities, contradictions and precarities. *Journal of Rural Studies*, *86*, 430–439. https://doi.org/10.1016/j.jrurstud.2021.07.009

Minkoff-Zern, L.-A. (2017). The case for taking account of labor in sustainable food systems in the United States. *Renewable Agriculture and Food Systems*, *32*(6), 576–578. https://doi.org/10.1017/S1742170517000060

Morgan, K., & Murdoch, J. (2000). Organic vs. Conventional agriculture: Knowledge, power and innovation in the food chain. *Geoforum*, *31*(2), 159–173. https://doi.org/10.1016/S0016-7185(99)00029-9

Morris, C. (2006). Negotiating the boundary between state-led and farmer approaches to knowing nature: An analysis of UK agri-environment schemes. *Geoforum*, *37*(1), 113–127. https://doi.org/10.1016/j.geoforum.2005.01.003

Moulton, A. A., & Popke, J. (2017). Greenhouse governmentality: Protected agriculture and the changing biopolitical management of agrarian life in Jamaica. *Environment and Planning. D, Society & Space*, *35*(4), 714–732. https://doi.org/10.1177/0263775816679669

Myers, N. (2015). Conversations on plant sensing: Notes from the field. *Nature and Culture*, *3*, 35–66.

Nally, D. (2011). The biopolitics of food provisioning. *Transactions - Institute of British Geographers (1965)*, *36*(1), 37–53. https://doi.org/10.1111/j.1475-5661.2010.00413.x

NCVO. (2022). *Employment status*. https://www.ncvo.org.uk/help-and-guidance/involving-volunteers/volunteers-and-the-law/employment-status/

Nettle, R. (2018). International trends in farm labour demand and availability (and what it means for farmers, advisers, industry and government). *International Agricultural Workforce Conference. Cork, Ireland*, 8–16.

NFU. (2017). *Vision for the future of farming: Access to a competent and flexible workforce*. https://www.nfuonline.com/archive?treeid=97783

NFU Horituclture and Potatoes Board. (2016). *The National Living Wage: A report to assess the potential financial implications for UK horticulture*. NFU. https://www.farminguk.com/content/knowledge/Assessing-the-impact-of-the-National-Living-Wage-on-horticultural-businesses(9748-9877-9312-9291).pdf

Ngutu, M., Bukachi, S., Olungah, C. O., Kiteme, B., Kaeser, F., & Haller, T. (2018). The actors, rules and regulations linked to export horticulture production and access to land and water as common pool resources in Laikipia County, Northwest Mount Kenya. *Land*, *7*(3). https://doi.org/10.3390/land7030110

Nicol, P., & Taherzadeh, A. (2020). Working Co-operatively for Sustainable and Just Food System Transformation. *Sustainability*, *12*(7). https://doi.org/10.3390/su12072816

Nieto, J. M. (2014). Labour and gender relations in Moroccan strawberry culture. In Jörg Gertel, & Sarah Ruth Sippel (Eds.), *Seasonal Workers in Mediterranean Agriculture* (pp. 217–228). Routledge.

Nuthall, P. L., & Old, K. M. (2018). Intuition, the farmers' primary decision process. A review and analysis. *Journal of Rural Studies*, *58*, 28–38. https://doi.org/10.1016/j.jrurstud.2017.12.012

Nye, C. & Lobley, M. (2021). *Farm labour in the U.K. Accessing the workforce the industry needs*. University of Exeter. http://socialsciences.exeter.ac.uk/media/universityofexeter/research/microsites/centreforruralpolicyresearch/pdfs/researchreports/Farm_labour_in_the_UK._Accessing_the_workforce_the_industry_needs_.pdf

O'Reilly, Karen & Rye, Johan Friedrik. (2021). The (re)production of the exploitative nature of rural migrant labour in Europe. In J. F. Rye & K. O'Reilly (Eds.), *International Labour Migration to Europe's Rural Regions* (pp. 228–245). Routledge.

OECD. (2010). *Learning for Jobs*. OECD. https://doi.org/10.1787/9789264087460-en

Office for National Statistics. (2019). *Labour in the agriculture industry, UK: February 2018*. https://www.ons.gov.uk/peoplepopulationandcommunity/populationandmigration/internationalmigration/articles/labourintheagricultureindustry/2018-02-06

OHCHR. (2020). *Statement by Ms. Hilal Elver, United Nations Special Rapporteur on the right to food, on her visit to Italy, 20-31 January 2020*. https://www.ohchr.org/en/statements/2020/01/statement-ms-hilal-elver-united-nations-special-rapporteur-right-food-her-visit?LangID=E&NewsID=25512

Oreszczyn, S., Lane, A., & Carr, S. (2010). The role of networks of practice and webs of influencers on farmers' engagement with and learning about agricultural innovations. *Journal of Rural Studies*, *26*(4), 404–417. https://doi.org/10.1016/j.jrurstud.2010.03.003

Osterman, P., Lowe, N., Anderson, B., Trotter, J. W., Iskander, N., & Agarwala, R. (2022). A Forum on the Politics of Skills. *ILR Review*, *75*(5), 1348–1368. https://doi.org/10.1177/00197939221110097

Palmer, J. (2021). Putting Forests to Work? Enrolling Vegetal Labor in the Socioecological Fix of Bioenergy Resource Making. *Annals of the American Association of Geographers*, *111*(1), 141–156.

Palumbo, L., & Sciurba, A. (2015). Vulnerability to Forced Labour and Trafficking: The case of Romanian women in the agricultural sector in Sicily. *Anti-Trafficking Review*, *5*, 89. https://doi.org/10.14197/atr.20121556

Papadopoulos, A., & Fratsea, L.-M. (2018). Migrant labour and intensive agricultural production in Greece: The case of the Manolada strawberry industry. In A. Corrado, C. de Castro, & D. Perrotta (Eds.), *Migration and Agriculture* (pp. 128–144). Routledge.

Patchett, M., & Mann, J. (2018). Five advantages of skill. *Cultural Geographies*, *25*(1), 23–29. https://doi.org/10.1177/1474474017739024

Payne, J. (2000). The unbearable lightness of skill: The changing meaning of skill in UK policy discourses andsome implications for education and training. *Journal of Education Policy*, *15*(3), 353–369. https://doi.org/10.1080/02680930050030473

Payne, J. (2003). *Choice at the end of compulsory schooling: A research review* (Research Report RR414). Department for Education and Skills.

Payne, J. (2017). The changing meaning of skill: Still contested, still important. In John Buchanan, David Finegold, Ken Mayhew, & Chris Warhurst (Eds.), *The Oxford Handbook of Skills and Training* (pp. 54–71). OUP.

Pedreño, A., Gadea, E., & De Castro, C. (2014). Labor, gender, and political conflicts in the global agri-food system: The case of the agri-export model in Murcia, Spain. In A. Bonanno & J. S. B. Cavalcanti (Eds.), *Research in Rural Sociology and Development* (Vol. 20, pp. 193–214). Emerald Group Publishing Limited. https://doi.org/10.1108/S1057-192220140000020008

Pereira, L. M.,, Drimie, S., Maciejewski, K., Tonissen, P. B., Biggs, R., & Environmental Governance. (2020). Food system transformation: Integrating a political–economy and social–ecological approach to regime shifts. *International Journal of Environmental Research and Public Health*, *17*(4), 1313. https://doi.org/10.3390/ijerph17041313

Pettenati, Giacomo, Toldo, Alessia, & Ferrando, Tomaso. (2019). The Food System as a commons. In J. L. Vivero Pol, Ferrando, Tomaso, de Schutter, Olivier, & Mattei, Ugo (Eds.), *Routledge Handbook of Food as Commons* (pp. 42–56). Routledge.

Phillips, H. (1985). Change in the UK glass-house industry and its impact on the rural environment. *The Industrialization of the Countryside. Geobooks, Norwich*, 121–143.

Piro, V., & Sanò, G. (2017). Entering the 'plastic factories': Conflicts and competition in Sicilian greenhouses and packinghouses. In A. Corrado, C. De Castro, & D. Perrotta (Eds.), *Migration and Agriculture* (pp. 293–308). Routledge.

Pitt, H. (2013). *Growing Together: An ethnography of community gardening as place making* [Cardiff]. https://orca.cardiff.ac.uk/id/eprint/53953/2/PittEH.pdf

Pitt, H. (2014). Therapeutic experiences of community gardens: Putting flow in its place. *Health & Place*, *27*, 84–91. https://doi.org/10.1016/j.healthplace.2014.02.006

Pitt, H. (2019). *Knowing to Grow: Working Paper 3 What is the problem with horticultural skills in the UK?* Cardiff University. https://www.cardiff.ac.uk/__data/assets/pdf_file/0008/1527848/Working-Paper-3-Problem-FINAL.pdf

Pitt, H. (2021a). *Keeping and improving seasonal workers*. Cardiff University. https://www.cardiff.ac.uk/sustainable-places/research/projects/knowing-to-grow#:~:text=Keeping%20and%20improving%20seasonal%20workers

Pitt, H. (2021b). What knowledge is required to grow food? A framework for understanding horticulture's skills 'crisis.' *Journal of Rural Studies*, *85*, 59–67. https://doi.org/10.1016/j.jrurstud.2021.05.001

Pitt, H. (2021c). *Building and owning as a team*. Cardiff University. https://www.cardiff.ac.uk/__data/assets/pdf_file/0009/2592324/GEOPL-Building-and-owning-as-a-team-November-2021-final.pdf

Polanyi, M. (1958). *Personal knowledge: Towards a post-critical philosophy*. Routledge & K. Paul.

Pollard, D. (2012). The gangmaster system in the UK: Perspective of a trade unionist. In *Ethical sourcing in the global food system* (pp. 115–127). Routledge.

Pollock, C. (2012). Repairing a fractured pipeline: Improving the effectiveness of agricultural R & D in the UK. *International Journal of Agricultural Management*, *2*(1), 1–4.

Preibisch, K. (2012). Migrant Workers and Changing Work-place Regimes in Contemporary Agricultural Production in Canada. *International Journal of Sociology of Agriculture and Food*, *19*(1), 62–82.

Press Association. (2004, May 18). Five years for death crash driver. *The Guardian*. https://www.theguardian.com/uk/2004/may/18/1

Previtali, F. S., & Fagiani, C. C. (2015). Deskilling and degradation of labour in contemporary capitalism: The continuing relevance of Braverman. *Work, Organisation, Labour & Globalisation*, *9*(1), 76–91. https://doi.org/10.13169/workorgalaboglob.9.1.0076

Pumares, P., & Jolivet, D. (2014). 'Origin Matters': Working Conditions of Moroccans and Romanians in the Greenhouses of Almería. In J. Gertel & S. R. Sippel (Eds.), *Seasonal Workers in Mediterranean Agriculture* (pp. 130–140). Routledge.

Pye, David. (1968). *The nature and art of workmanship*. Cambridge University Press.

Pye Tait Consulting. (2020). *Edible Horticulture Skills 2020*. AHDB. https://projectbluearchive.blob.core.windows.net/media/Default/Research%20Papers/Horticulture/CP%20189%20Skills%20surveys/1.%20Main%20report%20-%20all%20sectors.pdf

Raghuram, P. (2021). Interjecting the geographies of skills into international skilled migration research: Political economy and ethics for a renewed research agenda. *Population Space and Place*, *27*(5).

Reigada, A. (2017). Family farms, migrant labourers and regional imbalance in global agri-food systems. In A. Corrado, C. De Castro, & D. Perrotta (Eds.), *Migration and Agriculture* (pp. 95–110). Routledge.

Reigada, A., & de Castro, C. (2022). Agricultural labour in the global food system. In C. Sage (Ed.), *Research Agenda for Food Systems* (pp. 89–110). Edward Elgar Publishing.

Reyes-Garcia, V., Benyei, P., & Calvet-Mir, L. (2020). Traditional agricultural knowledge as a commons. In J. L. Vivero Pol, T. Ferrando, O. De Schutter, & U. Mattei (Eds.), *Routledge handbook of food as a commons*. Routledge.

Richards, T. J., & Rickard, B. (2020). COVID-19 impact on fruit and vegetable markets. *Canadian Journal of Agricultural Economics/Revue Canadienne d'agroeconomie*, *68*(2), 189–194. https://doi.org/10.1111/cjag.12231

Richardson, L., & Bissell, D. (2019). Geographies of digital skill. *Geoforum*, *99*, 278–286.
Robinson, G. M. (1983). The evolution of the horticultural industry in the vale of evesham. *Scottish Geographical Magazine*, *99*(2), 89–100. https://doi.org/10.1080/00369228308736554
Rogaly, B. (2008). Intensification of workplace regimes in British horticulture: The role of migrant workers. *Population, Space and Place*, *14*(6), 497–510. https://doi.org/10.1002/psp.502
Rogaly, B. (2021). Commentary: Agricultural racial capitalism and rural migrant workers. *Journal of Rural Studies*, *88*, 527–531. https://doi.org/10.1016/j.jrurstud.2021.07.006
Rogaly, B., & Qureshi, K. (2017). 'That's where my perception of it all was shattered': Oral histories and moral geographies of food sector workers in an English city region. *Geoforum*, *78*, 189–198. https://doi.org/10.1016/j.geoforum.2016.03.003
Rose, M. (Michael A.) (2005). The mind at work: Valuing the intelligence of the American worker. In *The mind at work: Valuing the intelligence of the American worker*. Penguin Books.
Rosset, P., & Val, V. (2020). The campesino a campesino agroecology movement in Cuba. In Jose Luis Vivero-Pol, Tomaso Ferrando, Olivier De Schutter, & Ugo Mattei (Eds.), *The Routledge Handbook of Food as Commons* (pp. 251–265). Routledge.
Rotz, S., Gravely, E., Mosby, I., Duncan, E., Finnis, E., Horgan, M., LeBlanc, J., Martin, R., Neufeld, H. T., Nixon, A., Pant, L., Shalla, V., & Fraser, E. (2019). Automated pastures and the digital divide: How agricultural technologies are shaping labour and rural communities. *Journal of Rural Studies*. https://doi.org/10.1016/j.jrurstud.2019.01.023
Rye, J. & Scott, S. (2021). Agricultural employers' representation and rationalisation of their work offer The 'benevolent moderator.' In J. F. Rye & O'Reilly, Karen (Eds.), *International Labour Migration to Europe's Rural Regions* (pp. 141–158). Routledge.
Rye, J. F., & Andrzejewska, J. (2010). The structural disempowerment of Eastern European migrant farm workers in Norwegian agriculture. *Journal of Rural Studies*, *26*(1), 41–51. https://doi.org/10.1016/j.jrurstud.2009.06.003
Rye, J. F., & Scott, S. (2018). International Labour Migration and Food Production in Rural Europe: A Review of the Evidence. *Sociologia Ruralis*, *58*(4), 928–952. https://doi.org/10.1111/soru.12208
Sage, C. (Ed.). (2022). *Research agenda for food systems*. EDWARD ELGAR.
SALT. (n.d.). *Solidarity Across Land Trades*. https://saltunion.wordpress.com/
Sanderson Bellamy, A., Furness, E., Nicol, P., Pitt, H., & Taherzadeh, A. (2021). Shaping more resilient and just food systems: Lessons from the COVID-19 Pandemic. *Ambio*, *50*(4), 782–793. https://doi.org/10.1007/s13280-021-01532-y
Saville, A., & Adams, A. E. (2021). Environmental justice in the American south: An analysis of black women farmworkers in Apopka, Florida. *Agriculture and Human Values*, *38*(1), 193–204. https://doi.org/10.1007/s10460-020-10147-0
Sbicca, J. (2018). *Food justice now! Deepening the roots of social struggle*. University of Minnesota press.
Schell, G. (2002). Farmworker Exceptionalism under the Law:How the Legal System Contributes to Farmworker Poverty and Powerlessness. In C. Thomspon & M. Wiggins (Eds.), *The Human Cost of Food*. University of Texas Press.
Schenner, J. K. (2018). The governance of the horticultural supply chain in the United Kingdom: A source of forced labour? *Economia Agro-Alimentare*, *20*(1), 29–54. https://doi.org/10.3280/ECAG2018-001003
Scott (2022). Scott, written evidence to 'Labour shortages in the food and farming sector' 2022, all written evidence available at https://committees.parliament.uk/work/1497/labour-shortages-in-the-food-and-farming-sector/publications/

Scott, S. (2013). Labour, migration and the spatial fix: Evidence from the UK food industry. *Antipode*, *45*(5), 1090–1109.
Scott, S., & O'Reilly, K. (2022). Media reporting of labour shortages in UK horticulture during the COVID-19 pandemic: The use of wartime metaphors in the selective unveiling of precarious work/workers. *Estudios Geográficos*, *83*(293), e109. https://doi.org/10.3989/estgeogr.202211 5.115
Scott, S., & Rye, J. F. (2021). Praised, prized, yet penalised: A critical examination of low-wage hiring queues in the global strawberry industry. *Journal of Rural Studies*, *88*, 473–481. https://doi.org/10.1016/j.jrurstud.2021.04.014
Scott, S., & Visser, M. A. (2022). Constraining labour: The integration dynamics of working-class horticultural migrants in rural areas of Norway, the UK and the US. *Sociologia Ruralis*, *62*(1), 112–130. https://doi.org/10.1111/soru.12363
Seabourne, C. (2023). *Tales from the Lenches*. Christopher Seabourne.
Selwyn, B. (2010). Globalized Horticulture: The Formation and Global Integration of Export Grape Production in North East Brazil. *Journal of Agrarian Change*, *10*(4), 537–563. https://doi.org/10.1111/j.1471-0366.2009.00245.x
Selwyn, B. (2014). Capital–Labour and State Dynamics in Export Horticulture in North-east Brazil. *Development and Change*, *45*(5), 1019–1036. https://doi.org/10.1111/dech.12113
Sennett, R. (2009). *The craftsman*. Penguin Books.
Sidwell, R. (1969). A Short History of Commercial Horticulture in the Vale of Evesham. *Vale of Evesham Historical Society Research Papers*, *2*, 43–51.
Siegmann, K. A., Quaedvlieg, J., & Williams, T. (2022). Migrant Labour in Dutch Agriculture: Regulated Precarity. *European Journal of Migration and Law*, *24*(2), 217–240. https://doi.org/10.1163/15718166-12340127
Sievers-Glotzbach, S., Euler, J., Frison, C., Gmeiner, N., Kliem, L., Mazé, A., & Tschersich, J. (2021). Beyond the material: Knowledge aspects in seed commoning. *Agriculture and Human Values*, *38*(2), 509–524. https://doi.org/10.1007/s10460-020-10167-w
Sigaut, F. (1993). Learning, Teaching, and Apprenticeship. *New Literary History*, *24*(1), 105–114. https://doi.org/10.2307/469273
Sijmonsa, A. (2022). It doesn't make sense to plant. *HortiDaily*. https://www.hortidaily.com/article/9401754/it-doesn-t-make-financial-sense-to-plant/
Sippel, S. R. (2014). Disrupted livelihoods? Intensive agriculture and labour markets in the Moroccan Souss. In J. Gertel & S. R. Sippel (Eds.), *Seasonal workers in mediterranean agriculture* (pp. 204–216). Routledge.
Sippel, S. R., & Gertel, J. (2014). Shared insecurities? Farmers and workers in Bouches-du-Rhone. In Jörg Gertel, & Sarah Ruth Sippel (Eds.), *Seasonal workers in mediterranean agriculture* (pp. 55–66). Routledge.
Skogstad, G. (1998). Ideas, Paradigms and Institutions: Agricultural Exceptionalism in the European Union and the United States. *Governance (Oxford)*, *11*(4), 463–490. https://doi.org/10.1111/0952-1895.00082
Soil Association. (2018). *Digging into horticulture: Encouraging the next generation of producers* [Policy Report]. Soil Association. https://www.soilassociation.org/media/15861/digging-into-horticulture.pdf
Sonn, J. W., Hess, M., & Wang, H. (2019). Spaces for creativity? Skills and deskilling in cultural and high-tech industries. *Geoforum*, *99*, 223–226. https://doi.org/10.1016/j.geoforum.2018.09.015
Sørensen, E. (2009). *The materiality of learning: Technology and knowledge in educational practice*. Cambridge University Press.
Sparrow, T. C. (2011). *Digging for a living: Market gardening in Badsey and Aldington*. Badsey Society.
Spiertz, J., & Kropff, M. (2011). Adaptation of knowledge systems to changes in agriculture and society: The case of the Netherlands. *NJAS-Wageningen Journal of Life Sciences*, *58*(1–2), 1–10.

Sporleder, T. L., & Moss, L. E. (2002). Knowledge management in the global food system: Network embeddedness and social capital. *American Journal of Agricultural Economics*, *84*(5), 1345–1352.

Stead, V. (2021). Precarity's reach: Intersections of history, life, and labour in the Australian horticultural industry. *The Journal of the Royal Anthropological Institute*, *27*(2), 303–320. https://doi.org/10.1111/1467-9655.13491

Steinberg, R. (1990). Social Construction of Skill: Gender, Power, and Comparable Worth. *Work and Occupations*, *17*(4), 449–482. https://doi.org/10.1177/0730888490017004004

Strauss, K. (2013). Unfree Again: Social Reproduction, Flexible Labour Markets and the Resurgence of Gang Labour in the UK. *Antipode*, *45*(1), 180–197. https://doi.org/10.1111/j.1467-8330.2012.00997.x

Strawbridge, C. L., Emmett, M. R., & Ashton, I. (2011). *Development of vocational training programmes with active links to current research* (Vol. 920). Scopus. https://www.scopus.com/inward/record.uri?eid=2-s2.0-84859716067&partnerID=40&md5=7094d877b622cf32d4d289b20c9aded1

Styles, G., Talks, I., & Tomlinson, H. (2022). *The Attraction of Agroecology and the barriers faced by new entrants pursuing agroecological farming and land work*. Landworkers Alliance. https://landworkersalliance.org.uk/wp-content/uploads/2018/10/Landworkers-Alliance-The-Attraction-of-AgroecologyFINAL.pdf

Sutherland, L.-A., & Marchand, F. (2021). On-farm demonstration: Enabling peer-to-peer learning. *The Journal of Agricultural Education and Extension*, *27*(5), 573–590. https://doi.org/10.1080/1389224X.2021.1959716

Taherzadeh, A. (2020). *Learning Pathways into Sustainable Agriculture*. Cardiff Universtity. https://blogs.cardiff.ac.uk/sustainableplaces/wp-content/uploads/sites/420/2020/01/Learning-Pathways-Into-Sustainable-Agriculture-Report.pdf

Thomas, R. J. (1986). The Mythology of Agricultural Exceptionalism: Some Comments. *In Defense of the Alien*, *9*, 18–21. JSTOR.

Timmermann, C., & Felix, G. F. (2015). Agroecology as a vehicle for contributive justice. *Agriculture and Human Values*, *32*(3), 523–538. https://doi.org/10.1007/s10460-014-9581-8

Tourte, L., & Gaskell, M. (2004). Horticultural auction markets: Linking small farms with consumer demand. *Renewable Agriculture and Food Systems*, *19*(3), 129–134. https://doi.org/10.1079/RAFS200475

Trewavas, A. (2003). Aspects of Plant Intelligence. *Annals of Botany*, *92*(1), 1–20. https://doi.org/10.1093/aob/mcg101

Tsing, A. (2012). Unruly Edges: Mushrooms as Companion Species: For Donna Haraway. *Environmental Humanities*, *1*(1), 141–154.

Underhill, E., & Rimmer, M. (2016). Layered vulnerability: Temporary migrants in Australian horticulture. *Journal of Industrial Relations*, *58*(5), 608–626. https://doi.org/10.1177/0022185615600510

Unwin, L. (2010). *Learning and Working from the MSx to the New Labour: Young People, Skills and Employment*. National Institute Economic Review 212. http://ner.sagepub.com.ezproxy.uwe.ac.uk/content/212/1/R49.full.pdf+html

USDA Economic Research Service. (2023). *Rural America at a Glance Economic Information Bulletin No. (EIB-246) 24 pp*. https://www.ers.usda.gov/publications/pub-details/?pubid=105154

Vallas, S. (1990). The Concept of Skill: A Critical Review. *Work and Occupations*, *17*(4), 379–398. https://doi.org/10.1177/0730888490017004001

van den Broek, D., Wright, C. F., Howe, J., & Reilly, A. (2021). Pro-market governance, migration status and worker vulnerability: The case of Australian horticulture. *Economic and Industrial Democracy*, *42*(4), 1305–1325. https://doi.org/10.1177/0143831X19861750

Veiga, Ubaldo Martínez. (2014). The political economy of El Ejido: Genealogy of the 2000 conflict. In J. Gertel & S. R. Sippel (Eds.), *Seasonal Workers in Mediterranean Agriculture* (pp. 103–111). Routledge.

Verdon, N. (2017). *Working the land a history of the farmworker in England from 1850 to the present day*. Palgrave Macmillan.

Verdon, N. (2019). Skill, Status and the Agricultural Workforce in Victorian England. *History (London)*, *104*(363), 829–850. https://doi.org/10.1111/1468-229X.12916

Vivero Pol, J. L. (2020). The idea of food as a commons' Multiple understandings for multiple dimensions of food. In J. L. Vivero Pol, T. Ferrando, De Schutter, O., & U. Mattei (Eds.), *Routledge handbook of food as a commons*. Routledge.

Vivero Pol, J. L., Ferrando, T., De Schutter, O., & Mattei, U. (2020). Introduction: The food commons are coming. In J. L. Vivero Pol, T. Ferrando, De Schutter, O., & U. Mattei (Eds.), *Routledge handbook of food as a commons*. Routledge.

Vivero Pol, J. L. (2017). The idea of food as commons or commodity in academia. A systematic review of English scholarly texts. *Journal of Rural Studies*, *53*, 182–201. https://doi.org/10.1016/j.jrurstud.2017.05.015

Walsh, J. (2014). From nations of immigrants to states of transience: Temporary migration in Canada and Australia. *International Sociology*, *29*(6), 584–606. https://doi.org/10.1177/0268580914538682

Wandersee, J. H., & Schussler, E. E. (1999). Preventing Plant Blindness. *The American Biology Teacher*, *61*(2), 82–86. https://doi.org/10.2307/4450624

Warhurst, C., & Gatta, M. (2017). A New Social Construction of Skill. In John Buchanan, David Finegold, Ken Mayhew, & Chris Warhurst (Eds.), *The Oxford Handbook of Skills and Training*. Oxford University Press.

Warhurst, C., Tilly, C., & Gatta, M. (2017). Introduction. Skills and Training: Multiple Targets, Shifting Terrain. In J. Buchanan, D. Finegold, K. Mayhew, & C. Warhurst (Eds.), *The Oxford Handbook of Skills and Training* (Vol. 1). Oxford University Press. https://doi.org/10.1093/oxfordhb/9780199655366.013.4

Weiler, A. M. (2022). Seeing the workers for the trees: Exalted and devalued manual labour in the Pacific Northwest craft cider industry. *Agriculture and Human Values*, *39*(1), 65–78. https://doi.org/10.1007/s10460-021-10226-w

Weiler, A. M., & Mclaughlin, J. (2019). Listening to migrant workers: Should Canada's Seasonal Agricultural Worker Program be abolished? *Dialectical Anthropology*, *43*(4), 381–388. https://doi.org/10.1007/s10624-019-09563-4

Weiler, A. M., McLaughlin, J., & Cole, D. C. (2017). Food Security at Whose Expense? A Critique of the Canadian Temporary Farm Labour Migration Regime and Proposals for Change. *International Migration*, *55*(4), 48–63. https://doi.org/10.1111/imig.12342

Weiler, A. M., Otero, G., & Wittman, H. (2016). Rock Stars and Bad Apples: Moral Economies of Alternative Food Networks and Precarious Farm Work Regimes. *Antipode*, *48*(4), 1140–1162. https://doi.org/10.1111/anti.12221

Wenger, E. (1998). *Communities of practice: Learning, meaning, and identity*. Cambridge University Press.

Wilbur, A. (2014). Cultivating Back-to-the-Landers: Networks of Knowledge in Rural Northern Italy. *Sociologia Ruralis*, *54*(2), 167–185. https://doi.org/10.1111/soru.12024

Wilkinson, J., & Goodman, D. (2019). Food Regime Analysis. In G. Allaire & B. Daviron (Eds.), *Ecology, Capitalism and the New Agricultural Economy* (pp. 142–162). Routledge.

Willett, W., Rockström, J., Loken, B., Springmann, M., Lang, T., Vermeulen, S., Garnett, T., Tilman, D., DeClerck, F., Wood, A., Jonell, M., Clark, M., Gordon, L. J., Fanzo, J., Hawkes, C., Zurayk, R., Rivera, J. A., De Vries, W., Majele Sibanda, L., … Murray, C. J. L. (2019). Food in the Anthropocene: The EAT–Lancet Commission on healthy diets from sustainable food systems. *The Lancet*, *393*(10170), 447–492. https://doi.org/10.1016/S0140-6736(18)31788-4

Worcestershire Local Enterprise Partnership. (2016). *Worcestershire Horticulture Strategy 2016—2025*. https://wlep.co.uk/wp-content/uploads/WLEP-Horticulture-Strategy-October-FINAL.pdf

Workaway. (n.d.). *How it Works*. https://www.workaway.info/en/info/how-it-works/host

WWOOF. (n.d.). *FAQs*. https://wwoof.org.uk/en/how-it-works

Zerbe, Noah. (2019). Food as Commodity. In J. L. Vivero Pol, Ferrando, Tomaso, De Schutter, Olivier, & Mattei, Ugo (Eds.), *Routledge Handbook of Food as Commons* (pp. 155–170). Routledge.

Index

Pages in *italics* refer to figures and pages in **bold** refer to tables.

FSC
www.fsc.org
MIX
Papier | Fördert gute Waldnutzung
FSC® C083411

Kolibri 360 GmbH
Ferdinand-Jühlke-Straße 7
99095 Erfurt, Deutschland
produktsicherheit@kolibri360.de